Genetic Factors in Atherosclerosis: Approaches and Model Systems

Monographs in
Human Genetics

Vol. 12

Series Editor
Robert S. Sparkes, Los Angeles, Calif.

Basel · München · Paris · London · New York · New Delhi · Bangkok · Singapore · Tokyo · Sydney

Genetic Factors in Atherosclerosis: Approaches and Model Systems

Volume Editors
Aldons J. Lusis, Los Angeles, Calif.
Robert S. Sparkes, Los Angeles, Calif.

25 figures and 22 tables, 1989

KARGER

Basel · München · Paris · London · New York · New Delhi · Bangkok · Singapore · Tokyo · Sydney

Monographs in Human Genetics

QH
431
M554
V.12

Library of Congress Cataloging-in-Publication Data
 Genetic factors in atherosclerosis.
 (Monographs in human genetics; vol. 12)
 Includes bibliographies.
 1. Atherosclerosis – Genetic aspects. 2. Lipoproteins – Metabolism – Disorders –
 Genetic aspects. I. Sparkes, Robert S., 1930– . II. Lusis, Aldons J., 1947– .
 III. Series: Monographs in human genetics; v. 12.
 [DNLM: 1. Arteriosclerosis – genetics. 2. Disease Models, Animal.
 W1 M0567P v.12/WG 550 G328]
 QH431. M554 vol. 12 [RC 692] 573.2′1 s 89-2666 [616.1′36042]
 ISBN 3–8055–4890–7

© Copyright 1989 by S. Karger AG, P.O. Box, CH–4009 Basel (Switzerland)
 Printed in Switzerland by Thür AG Offsetdruck, Pratteln
 ISBN 3–8055–4890–7

Contents

Preface

Atherosclerosis is a disease of the large arteries that is responsible for coronary artery disease and stroke, which, together, account for about half the deaths in Western populations. The single most significant rick factor in atherosclerosis is family history, implying a strong genetic component for the disease. The identification and characterization of the genes involved would have considerable clinical significance. First, it would undoubtedly provide insights into the molecular and cellular mechanisms responsible for the disease. Second, it could facilitate the development of new therapies as well as improve the targeting of therapies. Finally, it may even make it possible to predict which individuals are at high risk, allowing application of preventive measures.

Although certain relatively rare single gene defects, such as mutations of the low density lipoprotein receptor which occur in familial hypercholesterolemia, result in early onset atherosclerosis, the vast majority of disease is due to a combination of more subtle genetic variations and also environmental influences. How can these genetic factors be individually identified and studied on the noisy background of multiple genetic and environmental influences?

The most powerful strategy at present appears to be what has been called the 'candidate gene approach'. This involves: first, the identification of polymorphisms of genes likely to be involved in atherosclerosis (for example, genes influencing known risk factors for the disease, including cholesterol metabolism, hypertension, and diabetes); second, testing whether such polymorphisms are associated with the risk factors in populations or families; and third, characterizing the properties of the alleles of genes that are identified. Given the large number of genes likely to be involved and the important influences of poorly understood environmental factors (diet, cigarette smoking and exercise, for example), this will certainly be a formidable and long-term endeavor, but not an impossible one.

The problem can be simplified by subdivision into genetic influences affecting the various known risk factors. Which genes contribute to hyperlipidemias and other variations of lipoprotein metabolism associated with the disease? Which genes contribute to hypertension? Which genes contribute to diabetes? Within the various risk factor categories, it may be possible to define 'subphenotypes'. This will be of considerable importance in reducing the problem of genetic heterogeneity.

This volume focuses primarily on the role of cholesterol metabolism in atherosclerosis. The epidemiological and experimental evidence linking blood cholesterol with atherosclerosis, in both humans and animal models, is very strong. Chapter 1 by Angelo Scanu reviews this evidence, discusses lipoprotein phenotypes and subphenotypes associated with atherosclerosis, and summarizes current information on the diagnosis and treatment of lipoprotein disorders. From a clinical standpoint, it is of interest to note that, due to the fact that it is determined in large part by humoral factors, atherosclerosis, unlike most genetic disorders, is relatively accessible to dietary and drug therapy.

Among the key tools required for the identification of genes influencing risk factors in atherosclerosis are mathematical modeling approaches developed for analysis of data from nuclear families, extended pedigrees and twin studies. The uses of such methods, particularly as applied to lipoprotein variations, in both human populations and animal models, are described in Chapter 2 by Jean MacCluer. Such biometrical methodologies have provided strong evidence for the genetic determination of lipoprotein variations. Although the conclusions that have emerged from these analyses of lipoprotein variations have thus far been rather limited, they will certainly be extended considerably by the incorporation of data concerning the expression and segregation of various candidate genes.

In Chapter 3 we have compiled gene mapping results for various candidate genes involved in lipoprotein metabolism. We have constructed chromosomal 'fat maps' for both the human and mouse genes.

Chapter 4 by D. Galton and G. Ferns and Chapter 5 by Philippe Frossard and Sophia Vinogradov review approaches to the identification of genetic influences in atherosclerosis and summarize the results obtained to date. Chapter 4 focuses on candidate genes and Chapter 5 focuses on the identification of genes involved in complex disorders using DNA markers.

Human studies of genetic influences in atherosclerosis are complicated by environmental variables and by the difficulty of detailed biochemical and genetic analyses. These problems can be largely avoided by
the use of animal models. Moreover, experimental approaches in animals
allows the testing of hypotheses. Chapters 6–10 review the most important animal models for atherosclerosis.

Chapter 6 by Brian Van Lenten reviews the Watanabe Heritable
Hyperlipidemic (WHHL) rabbit, an animal model for familial hypercholesterolemia. The important information that has been obtained
from the WHHL rabbit clearly illustrates the advantages of animal models for both basic questions pertaining to lipid metabolism and clinical questions pertaining to mechanisms contributing to atherosclerosis.

Chapter 7 by Jan Rapacz and Judith Hasler-Rapacz reviews genetic
studies of atherosclerosis in swine. In particular, immunogenetic studies
carried out by these investigators and their collaborators over the past
two decades have led to the identification of a new genetic form of hypercholesterolemia associated with accelerated atherosclerosis. Unlike individuals with familial hypercholesterolemia and WHHL rabbits, the
affected swine have normal LDL receptor activity. Rather, the mutations
in swine appear to involve genetically altered lipoprotein particles, and
recent studies have revealed that mutations of apolipoproteins also contribute to hyperlipidemias in humans.

Chapter 8 by Kathy Laber-Laird and L. Rudel reviews genetic and
biochemical studies of hyperlipidemias and atherosclerosis in nonhuman primates. One criticism that has been made fo the use of animal
models of disease is that, due the metabolic and physiologic differences
between species, the conclusions drawn from animal studies may not
be applicable to humans. While this criticism does not appear to be just-
ified in general, nonhuman primates are very similar to humans in
terms of lipid metabolism and the nature of atherosclerotic lesions.
Moreover, they provide advantages in that environmental variables
can be controlled and that more detailed biochemical studies can be
performed.

Chapter 9 by Brian Ishida and Beverly Paigen reviews the mouse
model for atherosclerosis. The mouse is the classical mammal for genetic
studies and, as such, has several important and unique advantages for the
identification and characterization of genetic factors contributing to the
disease. This chapter puts to rest the notion that mice do not develop
atherosclerosis; it reviews recent studies supporting the concept that high
density lipoproteins protect against the disease; it summarizes information about various mutant strains of mice that may prove useful in studies

of lipoprotein metabolism and atherosclerosis; and it provides a current list of resources for researchers in the field.

In summary, this collection of reviews provides an overview of the most important approaches and model systems that promise to reveal the genetic influences contributing to atherosclerosis.

May, 1988   The Editors
                                                           Aldons J. Lusis, PhD
                                                           Robert S. Sparkes, MD

Lusis A J, Sparkes S R (eds): Genetic Factors in Atherosclerosis: Approaches and Model Systems. Monogr Hum Genet. Basel, Karger, 1989, vol 12, pp 1–49

Lipoprotein Disorders: Diagnosis and Treatment

Angelo M. Scanu

Department of Medicine and Biochemistry, University of Chicago, Pritzker School of Medicine and The Lipoprotein Study Unit and The Lipid Clinic, Chicago, Ill., USA

Historical Aspects

Although it was more than a century ago that a report appeared in the British literature describing, in human subjects, cutaneous fatty deposits referred to as xanthomas, the actual relationship between hyperlipidemia, xanthomas and premature coronary artery disease was not recognized until the 1930s and the genetic basis of familial hypercholesterolemia in the early 1950s [1] and placed on a more solid basis by Khachadurian [2] in 1964. The early observations stimulated interest in the study of plasma lipids and also led to a classification of human hyperlipidemias based on the knowledge of plasma cholesterol and triglyceride concentrations [3, 4]. The seminal investigations by Oncley [5] at Harvard in the late 1940s and by Gofman et al. [6] at Berkeley in the early 1950s provided new knowledge on the association of lipids with defined classes of plasma lipoproteins and opened up new insights into the understanding of lipid disorders which since then have been viewed as abnormalities of lipoprotein metabolism. The work by Gofman et al. [6] in particular, pointed to the association between elevated plasma levels of very low density lipoprotein (VLDL)[6] and low density lipoproteins (LDL), and a high incidence of cardiovascular disease (CVD). Unfortunately, a large-scale investigation, under the auspices of a federally sponsored cooperative study, failed to substantiate the Gofman finding. The conclusion by the panel that plasma cholesterol levels alone, independent of any knowledge of lipoprotein distribution, were a sufficiently good predictor of CVD, temporarily lessened the interest in plasma lipoprotein research [7].

A renewed interest in plasma lipoproteins in relation to hyperlipidemias and atherosclerosis arose from the studies of Fredrickson, Levy, and Lees [8]. These investigators, using quantitative lipid analyses and the separation of the major classes of serum lipoproteins by paper electrophoresis in an albuminated buffer, proposed a classification of the hyperlipoproteinemias which was later recognized by the World Health Organization and adopted by many clinical chemistry laboratories. The originally proposed five phenotypes were subsequently increased to six so that a distinction could be made between patients with only elevated plasma cholesterol (phenotype IIA) and those in whom hypercholesterolemia was associated with hypertriglyceridemia (phenotype IIB). However, as this classification was applied to an increasing number of subjects, its limits became apparent, particularly in terms of assessing the genetic basis of the phenotypes under consideration. This was clearly documented in a study of hyperlipidemic survivors of myocardial infarction, showing that the more commonly occurring phenotypes were the result of both genetic and nongenetic abnormalities [9]. As a consequence, a new classification of hyperlipidemias based on genetic studies was proposed [10, 11]. Subsequently the classic studies by Brown and Goldstein [12, 13] led to discovery that liver and other cells in the body contain specific membrane receptors for circulating LDL and that these receptors play a key role in cellular cholesterol homeostasis. This important discovery was followed by an uninterrupted series of exciting studies during which Brown and Goldstein and their associates elucidated the structure of the LDL receptor gene and provided a molecular basis for understanding familial hyperlipoproteinemia (FH). In the meantime, an impressive amount of new information was obtained on the properties of the main plasma apolipoproteins, most of which are now well defined in physical and chemical terms [14, 15], and their genes isolated and characterized. The application of molecular biology techniques to the study of plasma apolipoproteins has brought an important new dimension to the lipoprotein field and has permitted achievements previously deemed insurmountable. A remarkable one among them has been the elucidation of the complete structure of apo B_{100}, the apo B form of hepatic origin with a molecular weight in the range of 500,000 [16–22] and apo B_{48}, this apo B of intestinal origin [23, 24] and of apolipoprotein (a), a protein with a molecular weight varying between 300,000 and 1,000,000 exhibiting structural properties similar to human plasminogen [25, 26]. These new advances have permitted the identification of mutations both at the apolipoprotein level (charge and/or size differences) and the DNA level (restriction fragment length polymorphism, RFLP) and have made it possible to carry out systematic genetic analyses both at the individual and family levels. Such an informational growth also applies to three key enzymes in lipoprotein metabolism: lipoprotein lipase (LpL) [27], hepatic lipase

(HL) [28] and lecithin-cholesterol acyl transferase (LCAT) [29]. The structural details of the proteins involved in the transfer/exchange of cholesteryl esters and triglycerides in the plasma are also emerging.

This brief historical overview indicates that modern concepts concerning dyslipoproteinemias have evolved over the years from a series of basic studies on lipoprotein chemistry and metabolism. The current knowledge in this area is summarized in the following section.

Current Concepts in Lipoprotein Metabolism

Essentially all plasma lipids, which are water insoluble, are transported in the blood in association with special proteins, apolipoproteins, with which they form water-soluble complexes, lipoproteins [14, 15]. Lipoproteins are viewed as spherical or quasi-spherical particles having a core of either cholesteryl esters or triglycerides and a polar coat containing apolipoproteins, phospholipids and unesterified cholesterol. The general properties of the main plasma lipoproteins are given in table I, and those of the main apolipoproteins in table II. The main affiliations of each apolipoprotein with the major lipoprotein classes are summarized in figure 1.

Both intestine and liver are major contributors to lipoprotein production. In the so-called *exogenous pathway*, the intestine, during the process of absorption of exogenous fats, manufactures and secretes large triglyceride-rich particles, chylomicrons (diameter 80–500 mm) containing mainly apo B_{48}. Once these particles reach the circulation, they lose part of their triglyceride core through the concerted action at its cofactor apo C-II, in a reaction believed to occur at the endothelial surface of the capillaries of muscles and adipose tissue. As a result, chylomicron remnants are formed. These remnants are smaller than chylomicrons (diameter 30–50 mm), still contain apo B_{48} and by affiliating with plasma apo E they are taken up by the liver cells by receptor-mediated endocytosis. In the liver cells, the triglycerides of these remnants are degraded into fatty acids which are either oxidized or utilized for the synthesis of triglycerides and VLDL. The cholesteryl esters are cleaved in the lysosomes to free cholesterol which is then either converted to bile acids, used in membrane synthesis or incorporated into endogenous lipoproteins and secreted into the circulation.

In the so-called *endogenous pathway*, the liver synthesizes and secretes VLDL (diameter 40–80 nm). Some of the VLDL particles are destined to rapid uptake by the liver cells through the remnant and the LDL receptor pathways and some are degraded first to an intermediate density

Table I. The human plasma lipoprotein

Class	Diameter Å	Density g/ml	Electro-phoretic mobility	Major apolipo-protein	Triglycerides, % LM	Cholesteryl esters, % LM	Unesterfied cholesterol, % LM	Phospholipids, % LM	Proteins, % LM	Transport function
Chylo	800–5,000	0.93	origin	apo A-I, apo A-II, apo A-IV, apo B$_{48}$	86	3	2	7	2	dietary TG
VLDL	300–800	0.96–1.006	pre-β	apo B$_{100}$ apo Cs, apo E	55	12	7	18	8	endogenous TG
IDL	250–350	1.006–1.019	pre-β	apo B$_{100}$ apo E	23	29	9	19	29	endogenous TG and cholesteryl esters
LDL	210–230	1.019–1.063	β	apo B$_{100}$	6	42	8	22	22	endogeneous cholesteryl esters
HDL$_2$	100–120	1.003–1.125	α	apo A-I, apo A-II, apo Cs, apo E	5	17	5	33	40	cholesteryl esters from extrahepatic tissues
HDL$_3$	75–80	1.125–1.21	α	apo A-I, apo A-II, apo D	3	13	4	25	55	cholesteryl esters from extrahepatic tissues
Lp(a)	300–400	1.061–1.110	pre-β	apo B$_{100}$	3	33	9	22	33	unknown

% LM = percent lipoprotein mass.

Table II. The human plasma apolipoproteins

Apolipo-protein	Molecular weight	Origin	Lipoprotein affiliation	Function
Apo A-I	28,016	liver intestine	chylo, HDL	HDL structure, LCAT cofactor
Apo A-II	17,414	liver intestine	chylo, HDL	HDL structure? LCAT inhibitor overall, unknown
Apo A-IV	44,465	intestine	chylo, HDL	cofactor, LCAT
Apo B$_{48}$	264,000	intestine	chylo	chylomicron metabolism
Apo B$_{100}$	512,000	liver	VLDL, LDL	VLDL synthesis and secretion; LDL receptor recognition
Apo C-I	6,630	liver	chylo, VLDL, HDL	cofactor, LCAT?
Apo C-II	8,900	liver	chylo, VLDL, HDL	cofactor, LpL
Apo C-III	8,800	liver	chylo, VLDL, HDL	inhibitor of remnant uptake by liver
Apo D	22,000	liver	HDL	unknown
Apo E	34,135	liver	chylo, VLDL, IDL, HDL	binding to cell membrane receptors in liver (remnant and LDL receptor) and macrophages (LDL receptor)

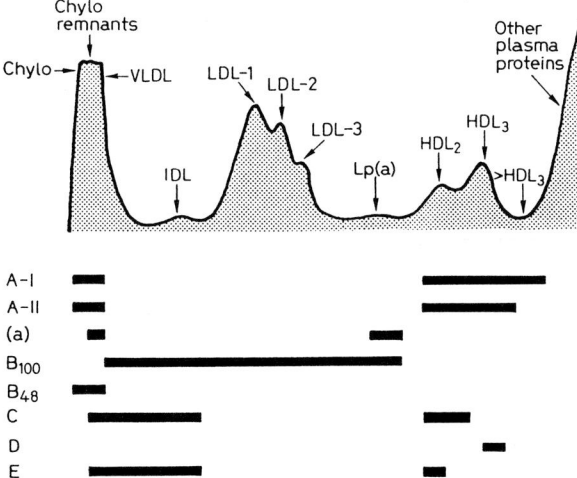

Fig. 1. Distribution of the main plasma lipoprotein classes of a normolipidemic human subject obtained by a single-step isopycnic density gradient ultracentrifugation technique. The apolipoprotein affiliation to each lipoprotein class is shown by solid bars.

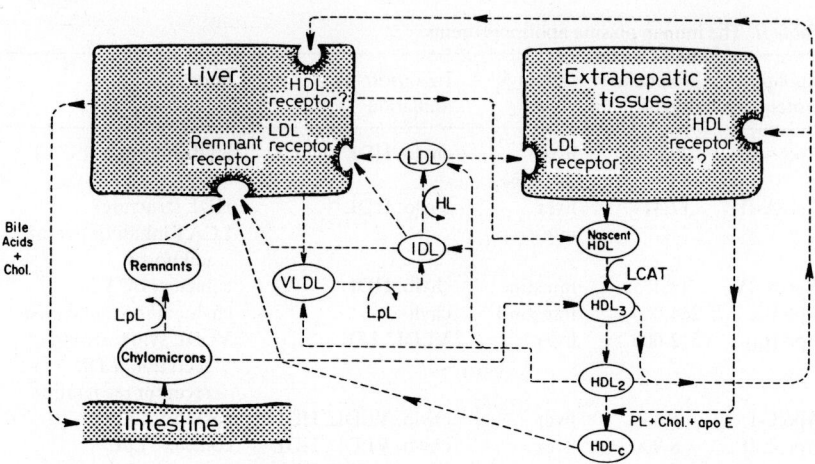

Fig. 2. Scheme of lipoprotein metabolism showing site of formation, intravascular degradation and tissue uptake of the main classes of plasma lipoproteins. Both exogenous and endogenous pathways are shown. For details see text.

lipoprotein (IDL) species (diameter 28–40 nm) and then to LDL through the progressive hydrolysis of the core triglycerides. Two major factors seem to determine the extent to which VLDL are converted to LDL: size and apo E content of VLDL and activity of hepatic LDL receptors [30]. The enzyme involved in the VLDL→ IDL conversion is LpL and its cofactor apo C-II. HL is involved in IDL→ LDL conversion and has no cofactor requirement.

The end product of the lipolytic cascade is LDL (diameter 18–28 nm) which is the major carrier of cholesterol in human plasma. LDL is taken up by the high affinity LDL receptor [31], which is an exquisite regulator of intracellular cholesterol homeostasis through a feedback mechanism involving the activity of hydroxymethylglutaryl-CoA (HMGCoA) reductase, a rate-limiting enzyme in cholesterol biosynthesis.

Besides the LDL receptor mechanism, intracellular cholesterol homeostasis is also influenced by HDL (diameter 5–12 nm) owing to the capacity of this lipoprotein class to act as a scavenger of cell cholesterol [14, 15]. This role, first proposed by Glomset [32], requires the concerted action of LCAT, which promotes the conversion of cholesterol to cholesteryl esters, and the cholesteryl estertriglyceride exchange/transfer protein system (CETP) [14, 15]. Cholesterol is transferred from HDL to both VLDL and LDL and ultimately degraded via the LDL receptor pathway. There are two additional ways by which HDL can contribute to reverse cholesterol transport. In one case, the largest of the HDL sub-classes, HDL_2, acquires apo E, becomes HDL_c (apo E-rich HDL) which is then taken up by the LDL receptor pathway. In the other case, HDL particles are directly taken up by the hepatocytes presumably via a high affinity site (HDL receptor?) which has not yet been fully characterized.

The important general concept that emerges from this outline is that lipoproteins, once secreted into the circulation, undergo intravascular remodeling resulting in products which have the necessary make-up for uptake and degradation by either the liver or peripheral cells. A scheme summarizing the major steps in lipoprotein metabolism is presented in figure 2. It may be noted that lipoprotein (a) (Lp(a)) is not included in the scheme because the physiological role of this LDL variant is yet un-known. Lp(a) is a genetically transmitted lipoprotein made of an LDL particle in which apo B_{100} is linked to apo(a) by disulfide linkage [33, 34]. Apo(a) is structurally similar to plasminogen and is made of isoform with molecular weights varying between 300,000 and 1,000,000 [25, 26].

General Comments on Diagnosis of Dyslipoproteinemias

The definition of hyper- or hypocholesterolemia requires knowledge of the 'normal' plasma cholesterol values. There is no universal agree-ment on what these normal values ought to be, although it is recognized that plasma cholesterol levels vary among populations and are affected by age, sex, type of diet and physical activity. Currently the most commonly adopted values are those derived from the Lipid Research Clinic Program [35] and from the recommendations by a Consensus Panel on Hyper-cholesterolemia summarized in table III [36]. Similarly, the values for normotriglyceridemia more commonly used are those recommended by the Consensus Panel on Hypertriglyceridemia [37].

When dealing with borderline values it should be taken into account that preanalytical variations can occur after the blood is drawn. In order to minimize these variations, blood is obtained after 12-hour fasting and is preferentially collected into tubes containing EDTA. In this way, clot-ting does not occur, plasma is rapidly separated from the blood cells so that lipoprotein changes due to oxidation may be avoided or minimized. Moreover, to prevent proteolytic events, blood is collected in tubes con-taining antiproteolytic agents such as aprotinin, kallikrein inhibitors, etc. However, no standard cocktail has yet been developed, and this is due to the fact that our knowledge of the proteolytic enzymes that have an action on plasma lipoproteins is yet limited. For patients at risk for CVD, the determination of plasma cholesterol and triglycerides may be

Table III. Plasma lipid and lipoprotein cholesterol concentrations (mg/dl)

Age, years	Plasma cholesterol			Plasma triglyceride			LDL cholesterol			HDL cholesterol		
	percentiles			percentiles			percentiles			percentiles		
	10	50	90	10	50	90	10	50	90	10	50	90
White males												
15–19	120	146	183	43	69	120	68	93	123	34	46	59
20–24	130	165	204	50	86	165	73	101	138	32	45	57
25–29	143	178	227	54	95	199	75	116	157	32	44	58
30–34	148	190	239	58	104	213	88	124	166	32	45	59
35–39	157	197	249	62	113	251	92	131	176	31	43	58
40–44	163	203	250	64	122	248	98	135	173	31	43	60
45–49	169	210	258	68	124	253	106	141	186	33	45	60
50–54	169	210	261	68	124	250	102	143	185	31	44	58
55–59	167	212	262	67	119	235	103	145	191	31	46	64
60–64	171	210	259	68	119	235	106	143	188	34	49	69
65–69	170	210	258	64	112	208	104	146	199	33	49	74
70+	162	205	252	67	111	212	100	142	182	33	48	70
White females												
15–19	126	154	190	44	66	107	67	93	127	38	51	68
20–24	130	160	203	41	64	112	62	98	136	37	50	68
25–29	136	168	209	42	65	116	73	103	141	40	55	73
30–34	139	172	213	44	69	123	76	108	142	40	55	71
35–39	147	182	225	46	73	137	81	116	161	38	52	74
40–44	154	191	235	51	82	155	89	120	164	39	52	78
45–49	161	199	247	53	87	171	90	127	173	39	56	78
50–54	172	215	268	59	97	186	102	141	192	40	59	77
55–59	183	228	282	63	106	204	103	148	204	39	58	82
60–64	186	228	280	64	105	202	105	151	201	43	60	85
65–69	183	229	280	66	112	204	104	156	208	38	60	79
70+	180	226	278	69	111	204	107	146	189	37	60	82

insufficient to establish the nature of the lipid disorder. Integrated chemical, ultracentrifugal, and immunological analyses may be needed to gain information on the distribution of cholesterol among the main classes of plasma lipoproteins, and on the total plasma levels of the main apolipoproteins such as apo A-I, apo B inclusive of Lp(a). Extensive laboratory analyses should prove particularly rewarding in subjects with borderline hypercholesterolemia because of the likelihood to uncover either high plasma Lp(a) levels, low HDL, or a dissociation between LDL cholesterol and apo B. The knowledge of the apo E isoform patterns is also useful particularly in cases where hypercholesterolemia is associated with hypertriglyceridemia such as in type III dysbetalipoproteinemia.

The Working of a Modern Lipid Clinic

Clinic

A successful Lipid Clinic should have: (1) one or more qualified physicians with a good understanding of the basic principles of lipoprotein metabolism and experience in the diagnosis and treatment of dyslipoproteinemias; (2) a dietician with a direct experience in the treatment of these patients; (3) a dedicated clinical coordinator; (4) a dependable laboratory able, when necessity arises, to provide specialized tests (see below); (5) access to consultants in areas of high risk of CVD, i.e., diabetes, hypertension, heart and peripheral vascular disorders. An ideal Lipid Clinic should also have a pediatric component particularly in view of the fact that genetic abnormalities are often observed early in life. Moreover, the children–adolescent population is ideal for implementing preventive measures and educational programs.

Laboratory

In principle, access to the highest number of analyses should help in better identifying the nature of the metabolic abnormality or abnormalities present in patients with dyslipoproteinemia. However, the provision of an extensive battery of lipoprotein and apolipoprotein assays on a service basis is made difficult by the lack of standardization of many of the techniques currently used, by their cost and also by the limited awareness of the significance of the results by the average practitioner. Overall, it is best to adopt laboratory analyses to the perception of needs by the physician and to the type of laboratory available. The usefulness of total plasma cholesterol, HDL cholesterol and total plasma triglycerides is well recognized. On the other hand, such a usefulness is less apparent for plasma Lp(a) and apolipoproteins in general also because these assays are comparatively less available. Thus, a case can be made for a specialized laboratory able to provide state-of-the-art technology. For example, at the University of Chicago, the Lipid Clinic has ready access to a dedicated Lipoprotein Laboratory which operates on a service basis. In what is referred to as 'Lipoprotein Profile' (fig. 3) the major plasma lipoprotein classes are separated by a single-step isopycnic density gradient ultra-

The University of Chicago

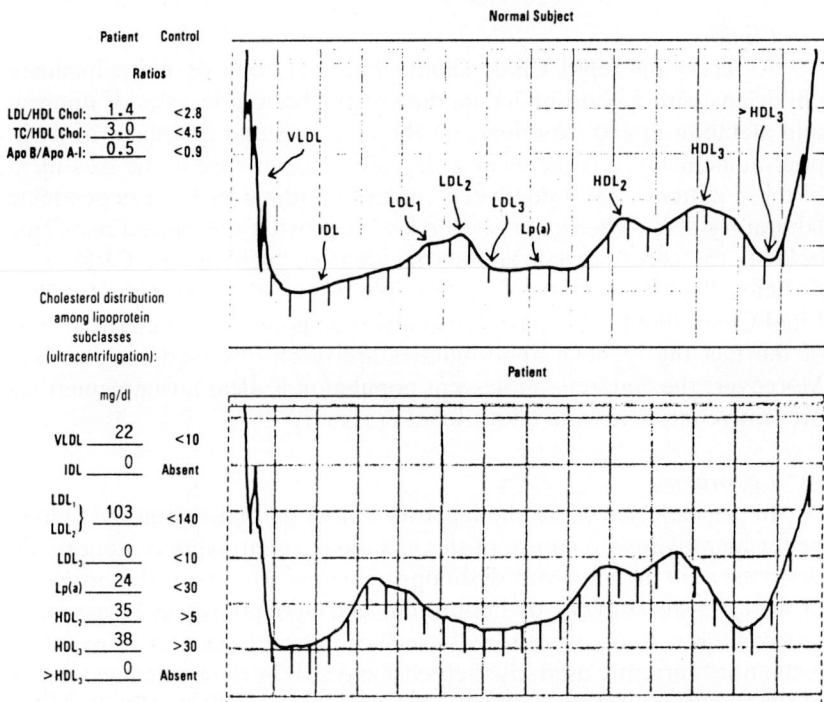

LIPOPROTEIN PROFILE and APOLIPOPROTEIN VALUES

L.F.   **34 y.o. Female**

Referring Physician: **Dr. L.R.**   **Med. I.**   **10-1-87**

Cholesterol/Triglyceride Values		Normal Range	Apolipóprotein Values		Normal Range
Total Cholesterol:	222	mg/dl (150-210 mg/dl)	Apo A-I:	168	mg/dl (100-140 mg/dl)
Total Triglycerides:	92	mg/dl (70-150 mg/dl)	Apo B:	83	mg/dl (90-110 mg/dl)
HDL Cholesterol, total:	73	mg/dl (35-60 mg/dl)	Others:		mg/dl
LDL Cholesterol, total:	103	mg/dl (70-159 mg/dl)	Apo(a) isoforms:		
			Apo E isoforms:	3/3	

Normal Subject

	Patient	Control
		Ratios
LDL/HDL Chol:	1.4	<2.8
TC/HDL Chol:	3.0	<4.5
Apo B/Apo A-I:	0.5	<0.9

VLDL, IDL, LDL₁, LDL₂, LDL₃, Lp(a), HDL₂, HDL₃, >HDL₃

Cholesterol distribution among lipoprotein subclasses (ultracentrifugation):

Patient

	mg/dl	
VLDL	22	<10
IDL	0	Absent
LDL₁ LDL₂	103	<140
LDL₃	0	<10
Lp(a)	24	<30
HDL₂	35	>5
HDL₃	38	>30
>HDL₃	0	Absent

Interpretation:

Total plasma cholesterol above the 90th percentile. This mild hyper-cholesterolemia is due to both Lp(a) and high HDL (see values of HDL cholesterol and apo A-I). The normal values of LDL cholesterol and apo B are compatible with this interpretation. Note normal ratio parameters.

5/87 AMS: __MG__ : __BK__   Signature: _a.m.scanu_
Reference #: __3669__   Director, Lipoprotein Laboratory

Fig. 3. Example of a 'lipoprotein profile', a report integrating data obtained by a single-step density gradient ultracentrifugation method with chemical analyses and immuno-quantification of apo B and apo A-I. This report is obtained before a patient is seen at the Lipid Clinic of the University of Chicago. Isoform analyses of apo E and apo(a) are performed if the need is perceived by the examining physician.

centrifugation which permits the identification of LDL polymorphs, Lp(a), HDL_2 and HDL_3 [38]. This analysis is complemented by chemical determinations of total plasma cholesterol, HDL cholesterol and total triglycerides from which the value of LDL cholesterol is derived and by the immunological quantitation of total plasma apo A-I and apo B. Isoform analyses for apo E and apo(a) by electrophoretic techniques are also provided. Referring physicians not familiar with lipid metabolism and lipoprotein disorders are usually unable to readily interpret such an assembly of relatively sophisticated analyses. Thus, it is important to provide a detailed interpretation of the results together with a suggested diagnosis to help in the overall clinical evaluation. Our experience indicates that the success of a Lipid Clinic depends on the ability of the laboratory to provide quality control data and respond to requests for analyses less readily available such as quantification of both apo B_{100} and apo B_{48}, measurement of the activities of LCAT and of post-heparin LpL and HL. Moreover, cDNA and genomic clones of apolipoproteins are now available in some major centers and, whenever indicated, appropriate RFLP analyses may be carried out. It would also be ideal to have access to facilities for obtaining skin biopsies and to grow fibroblasts in culture for patients where a deficiency of the LDL receptor is suspected. Finally, in difficult cases, the services of a Clinical Research Center would permit a close observation of patients under given dietary protocols and the performance of dynamic studies not possible on an outpatient basis.

Background Information on the Treatment of Dyslipoproteinemas

Diet

The first approach to the treatment of lipoprotein disorders is diet; drugs, if needed, are added to the dietary regimen but do not replace it [39]. The following basic principles apply to all forms of hyperlipoproteinemia: (1) achieve and maintain ideal body weight; (2) adhere to dietary principles as close as possible even when away from home; (3) obtain dietary advice from experienced dieticians able to establish a good rapport with the patient. The selection of a diet depends on the nature of the hyperlipoproteinemia and requires knowledge of the mechanism of action of each nutrient on plasma lipoproteins although in some cases the available information may not be as detailed as needed. The general properties of the dietary components used in the treatment of hyperlipoproteinemias are outlined below.

Cholesterol

Ingestion of cholesterol can affect the levels of plasma cholesterol although to a different extent in various individuals. An average American adult ingests 400–500 mg/day cholesterol while about twice as much enters the intestine via the bile. Evidence indicates that from 40 to 60% of the cholesterol entering the intestine is absorbed. According to Grundy [39], this limitation in absorptive capacity may protect human subjects from developing high plasma levels of cholesterol even after ingestion of relatively large amounts of cholesterol. A prudent diet providing an average of 300 mg cholesterol/day is the one commonly advised for patients with hypercholesterolemia. A more stringent diet may not be necessarily more efficient and difficult to comply with for a prolonged period of time.

Saturated Fatty Acids

These fatty acids are known to cause elevation of plasma cholesterol and to particularly raise plasma LDL. The most common of these are: lauric acid (12:0), myristic acid (14:0), plamitic acid (16:0) and stearic acid (18:0). Their action appears to be directed at reducing the clearance of LDL from the circulation [39], presumably by reducing the activity of the LDL receptor. Saturated fats may raise the levels of plasma VLDL and to a relatively small extent plasma HDL. In view of the recognized atherogenicity of LDL, patients with hypercholesterolemia should be advised to reduce their intake of saturated fats with preference for polyunsaturated and monounsaturated fats.

Polyunsaturated Fatty Acids

Polyunsaturated fatty acids have seen extensive use in the treatment of hypercholesterolemia because, when substituted for saturated fatty acids in a diet, they lower plasma cholesterol levels. Typical representatives are linoleic acid (18:2w6) which is an essential fatty acid involved in the synthesis of prostaglandins. How linoleic acid or other polyunsaturated fatty acids lower plasma cholesterol is not precisely known. Suggested mechanisms are: (1) increased fecal excretion of cholesterol; (2) changes in lipoprotein structure; (3) redistribution of cholesterol among body tissues. Since the hypocholesterolemic action of linolenic acids is at the expense of LDL [37], one may speculate that this action is either at the levels of the synthesis of apo B_{100} or an increased activity of the LDL receptor. Polyunsaturated fatty acids do also lower plasma HDL [40], apparently by decreasing the synthesis and secretion of apo A-I, the major component of HDL [41, 42]. Low levels of plasma HDL are believed to represent a risk factor for cardiovascular disease [43, 44]. In this context, polyunsaturated fatty acids may be considered harmful. It should also be noted that linolenic acid in high dosage can increase the incidence of tumors in experimental animals, be immunosuppressive and increase the incidence of cholesterol gallstones [39].

Polyunsaturated fatty acids which recently have received a great deal of attention are the ones found in fish, particularly from cold ocean waters. Fish do not synthesize essential fatty acids of the w-3 type but acquire them by eating the ocean plankton. These acids contain from 16 to 22 carbon atoms and from 3 to 6 double bonds which are formed by elongation and desaturation of linolenic acid. In w-3 fatty acids the first double bond is in position 3 from the terminal methyl group. In turn, polyunsaturated fatty acids from vegetable source are w-6, because the first double bond is in position 6 from the methyl group. The w-3 fatty acids are the precursor of the E_3 series of prostaglandins, are incorporated into the phospholipids of cell membranes preferentially over triglycerides and decrease the synthesis and secretion of VLDL [45]. They may also induce changes in the lipid composition of platelet membranes and thus are responsible for the inhibitory effect of w-3 fatty acids on platelet aggregation. Fish oils, particularly those rich in w-3 fatty acids, may be used in the treatment of hypertriglyceridemias. Currently, there is an uncertainty about the optimal therapeutic dosage of these oils. Whereas diets that include fish servings 3–4 times per week may have a preventive role for CVD, the use of fish oil extracts as dietary supplements is still open to question in terms of efficacy, quality control of products and long-term safety. At this time, the liberal use of these oils without medical supervision should be discouraged.

Monounsaturated Fatty Acids

Olive oil (16:1w6) is the more common representative of the monounsaturated fatty acids. It is contained in olive and rapeseed oil and has recently been shown to have a hypocholesterolemic effect comparable to that of polyunsaturated fats [39]. Oleic acid does not lower the plasma levels of HDL and has been an integral part of the Mediterranean diet for several generations with no known side effects. Moreover, oleic acid is normally synthesized by the body and, being a monounsaturated fatty acid, is less likely to undergo oxidation than polyunsaturated fats [39]. Olive oil appears to have all of the attributes to become a regular component of the average American diet. However, there is a need for educational efforts in this regard and a more competitive market to reduce product cost.

Protein

Our knowledge in this area is rather limited. Some studies have suggested that proteins from vegetable origin like soy protein have a hypocholesterolemic effect. Grundy and Abrams [46] have failed to confirm these findings and have suggested that if plant and animal proteins have a different effect on plasma cholesterol levels, the differences are likely to be small. On the other hand, the same authors have observed that soy protein lowers plasma triglyceride levels in hypertriglyceridemic subjects [46]. The mechanism remains to be elucidated.

Carbohydrates

Three types of dietary carbohydrates can affect plasma lipid levels and lipid metabolism in general: simple sugars, i.e. mono- and disaccharides, complex carbohydrates which are digested to simple sugars in the intestine, and complex carbohydrates which are not amenable to digestion. Diets high in carbohydrates increase both synthesis and secretion of VLDL [47, 48]. Simple sugars are believed to be more effective than complex sugars in this regard but this notion rests on limited experimental evidence. The degree of hypertriglyceridemic response to a high carbohydrate diet varies from individual to individual; compared to normal subjects it is higher in patients with congenial hypertriglyceridemias and lag time onset is also comparatively shorter. High carbohydrate diets also tend to lower the plasma HDL levels. Overall, it is difficult to assess the effect that simple and digestible complex sugars have on lipoprotein metabolism since it depends on the metabolic state of the individual and on the other dietary components, each of which may have independent influence on the plasma lipid levels. The average Mediterranean diet which is rich in carbohydrates, low in animal fats and flavored by olive oil has been credited for a low incidence of CVD. However, 'genetic factors' and general living conditions might also be determinants in maintaining low risk. In terms of indigestible carbohydrates, their presmptive hypocholesterolemic action has stimualted the market to produce fiber products particularly as ingredients of cereals. Several types of fibers are known: cellulose, pectins, gums, etc., all complex polysaccharide polymers which are not attacked by the digestive enzymes. A reasonable documentation is available for the hypocholesterolemic action of pectin which consists of polymers of galacturonic acid with pentose and hexose side chairs [39]. A hypocholesterolemic action has also been attributed to gums, complex polysaccharides containing glucuronic and galacturonic acid, xylose, arabinose and mannose. However, more systematic studies are needed on the mechanism of action, dosage and safety of dietary fibers before they see common use in the treatment of hyperlipoproteinemias.

Alcohol

Alcohol is a common component of the American diet consumed both at meal times and/or on a social level. It is often difficult to quantify alcoholic intake. Usually people are classified as moderate, intermediate or heavy drinkers. In terms of effects of plasma lipoprotein levels there is an individual variation in response. However, such a response is marked in those with an underlying hyperlipidemias. In these cases, a marked increase in the production and secretion by the liver of triglyceride-rich particles is observed. In general, dietary alcohol induces an increase in total plasma triglyceride levels and VLDL triglycerides and also an increase in the HDL_3 subclass but not HDL_2 [39]. Since HDL_2 but not HDL_3 appears to be inversely related to the incidence of CVD, alcohol may not be viewed as preventive for CVD and, in fact, potentially atherogenic. In heavy drinkers, the effect of alcohol is complicated by its potential toxic action on the liver as reflected by a relative deficiency of LCAT and, in more advanced stages, decreased cholesterol and lipoprotein synthesis. Whether there is a virtue in asking people who do not drink alcohol to add it to their diet is questionable. In our personal experience it is best to keep patients away from alcohol in order to avoid possible addiction.

General Dietary Principles

Hypercholesterolemia. The diet recommended by the American Heart Association (AHA) consists of a three-phase plan (table IV) where the intake of total fat, saturated fatty acids and cholesterol are progressively reduced according to the severity of hypercholesterolemia and degree of responsiveness [49].

Because of the potential adverse effects, polyunsaturated fatty acids may be replaced by monounsaturated fats such as those contained in olive oil under appropriate dietary supervision. The response of each phase of the AHA diet must be assessed at least for 3 months before a decision is made to either increase dietary restrictions or add to the diet drug therapy. Phase 3 is rather strict and may be difficult to follow. In such extremes and when the ideal hypocholesterolemic response is not achieved, it is best to add drug therapy to the phase 2 diet.

Hypertriglyceridemias With or Without Hypercholesterolemia. The most common forms of this type of dyslipoproteinemias are those due to either overproduction of VLDL, decreased degradation, or both. In obese patients, the first approach is weight loss to be achieved through restriction of total calories and carbohydrate intake particularly of simple sugars. The decrease of the plasma triglycerides and VLDL levels usually

Table IV. AHA recommended diet

Phase I
Composition
> Limit fat intake to 30% of total calories with a sat:mono:poly ratio of 1:1:1, dietary cholesterol less than 300 mg/day.

General Description
> Limit meat to no more than 7 oz a day.
>> Use fish and poultry more frequently than other meats. Include only chicken and turkey with skin removed. Any fatty fish (e.g., salmon) is acceptable for Phase I. Use lean cuts of veal, beef, pork, or lamb.
>
> Restrict whole eggs to two per week, including those used in cooking (egg white may be used as desined).
> Restrict milk products to 1% fat milk, ice milk, sherbert, low-fat frozen yogurt, low-fat cheese, and low-fat cottage cheese.
> Avoid hard fats such as butter, regular cheeses, lard, coconut oil, palm oil, chocolate candy; for fats, use only vegetable oils, olive oil, or soft-tub margarines.
> Bread, cereals, pasta, potatoes, and rice are allowed, except when made with egg yolks.
> Avoid whole-milk products, marbled meats, fish eggs, organ meats, bakery goods made with hard fats and egg yolks, and rick desserts.

Phase II
Composition
> Limit fat to 25% of total calories with a sat:mono:poly ratio of 1:1:1. Dietary cholesterol less than 200 mg/day.

General Description
> Limit meat to no more than 6 oz a day.
>> Restrict red meat in favor of fish and poultry. Use chicken and turkey with skins removed, and use lean cuts of meat only.
>
> Eat no egg yolks, although egg whites and egg substitutes are allowed.
> Restrict milk products to 0.5% fat milk, ice milk, sherbert, low-fat cheese, and low-fat cottage cheese.
> Avoid hard fats such as butter, regular cheeses, lard, coconut oil, palm oil, and chocolate candy; for fats, use only vegetable oils, olive oil, or soft-tub margarines.
> All vegetables and fruits are allowed except for coconut; limit olives and avocados.
> Bread, cereals, pasta, potatoes, and rice are allowed, except when made with egg yolks; limit starchy foods to prevent weight gain.
> Avoid whole-milk products, marbled meats, fish eggs, organ meats, bakery goods made with hard fats and egg yolks, and rich desserts.

Phase III
Composition
> Limit fat to 20% of total calories with a sat:mono:poly ratio of 1:1:1. Dietary cholesterol less than 150 mg/day.

General Description
> Limit meat to 3 oz a day. Restrict red meat in favor of fish and poultry. Use chicken and turkey with skins removed, and use lean cuts of meat.

Table IV. (Continued)

Use no egg yolks, but egg whites and egg substitutes are allowed.

Restrict milk products to skim milk, skim milk yogurt, and cheese with less than 1% fat.

Avoid hard fats; for fat use only small quantities of vegetable oils, olive oil, or soft-tub margarines.

All vegetables are allowed except for coconut, olives, and avocados.

Cereals, pasta, potatoes, rice and fat-free breads are allowed, but not if made with egg yolks.

Avoid whole-milk products, marbled meats, fish eggs, nuts, organ meats, bakery goods made with hard fats and egg yolks, and rich desserts.

parallels weight loss and one may expect that upon achieving ideal body weight, normalization of the plasma lipid levels will also ensue. However, there are cases where mild to severe hypertriglyceridemias are not associated with obesity. Eucaloric diets should be recommended and carbohydrate intake reduced to about 30% of the total calories. A regular exercise program should be adjunct to the dietary regimen after an appropriate evaluation of the cardiac function. Such an exercise program promotes utilization of substrate for energy by peripheral tissues thus preventing the liver from producing triglyceride-rich particles.

In severe exogenous hypertriglyceridemias characterized by high level of plasma chylomicrons, the intake of dietary fats must be drastically curtailed to less than 10% of the total caloric intake and medium-chain triglycerides may replace the long-chain triglycerides. In these cases, obesity is not necessarily present and weight loss does not usually bring a significant improvement of the hyperlipoproteinemia.

Drugs

In this section, the main characteristics of the drugs most commonly used in the treatment of lipoprotein disorders will be listed. Their specific use will be discussed when dealing with each individual disease state. Recent reviews on the subject have appeared [50–53].

Bile Acid – binding Resin (Cholestyramine and Colestipol)

Either cholestyramine (Questran; Mead Johnson) or colestipol (Colestid; Upjohn Co.) may be used in the treatment of hypercholesterolemias that have responded poorly to a dietary regimen. These agents lower plasma LDL cholesterol and apo B and their effect is usually dose dependent.

Chemistry. The structural formulas of cholestyramine and colestipol are given below. Cholestyramine is the chloride salt of a basic ion-exchange resin in which the ion-exchange sites are provided by trimethylbenzylammonium groups in a large copolymer of styrene and divinylbenzene. Colestipol is a copolymer of diethylpentamine and epichlorohydrin. Both of these agents are insoluble in water, are not degraded by the digestive enzymes and are not absorbed by the intestine.

$$\left[\begin{array}{c} ...-CH-CH_2-CH-CH_2-... \\ \bigcirc \quad \bigcirc \\ ...-CH_2-CH-... \quad CH_2N^+(CH_3)_3 \; Cl^- \end{array} \right]_n$$

Cholestyramine

$$\left[\begin{array}{c} HNCH_2CH_2\,NCH_2CH_2NCH_2CH_2NCH_2CH_2NH \\ CH_2 \quad CH_2 \quad CH_2 \quad CH_2 \quad CH_2 \\ HCOH \quad HCOH \quad HCOH \quad HCOH \quad HCOH \\ CH_2 \quad CH_2 \quad CH_2 \quad CH_2 \quad CH_2 \\ -CH_2CH_2NCH_2CH_2N \quad HNCH_2CH_2N \quad HNCH_2CH_2- \end{array} \right]_n$$

Colestipol

Mechanism of Action. These resins bind bile acids in the intestine thus preventing their absorption and causing their fecal excretion coupled with that of cholesterol. As a consequence, the liver hydroxylase is free to promote the intrahepatic conversion of cholesterol to bile acids. The net result is the loss of bile acids and cholesterol from the liver cells which compensate for these losses by increasing the number of LDL receptors and an increase in the activity of HMGCoA reductase, a rate-limiting enzyme in cholesterol synthesis. This ultimately leads to a rise in the endogenous production of cholesterol by the liver cells. Moreover, the increase in the number of LDL receptors causes a greater uptake of LDL particles from the plasma and a decrease in the levels of these lipoproteins in the plasma. Thus, it is apparent that LDL receptors are not only involved in the mechanism of familial hypercholesterolemia but are also necessary for the hypocholesterolemic action of bile acid-binding resins. This provides an explanation for the known ineffectiveness of bile acid-binding resins in homozygous familial hypercholesterolemia.

Indications. Bile-acid sequestrants are most effective in either familial or polygenic hypercholesterolemia both characterized by elevated levels of plasma LDL. The drop in LDL levels is dependent on drug dosage usually from 12 to 24 g to be taken 3 times daily within 1 h from meals. To prevent interference with their absorption the intake of other drugs is avoided 1 h before and 4 h after the oral administration of the resin. The decrease in the plasma levels of LDL is maximal within 2 weeks and it is retained with the continuation of the therapy. There is no major effect on the plasma levels of HDL. Neither cholestyramine nor colestipol is included in the treatment of endogenous hypertriglyceridemias because these resins tend to significantly increase the levels of VLDL and IDL. A rise in plasma triglycerides is also seen in hypercholesterolemic patients treated with these resins. However, this hypertriglyceridemia is usually mild and may disappear in a few weeks from the initiation of the therapy.

Side Effects. The most common complaints are nausea, indigestion, abdominal discomfort and constipation. The latter can be favorably influenced by the use of bran cereals. At high doses of resin, patients may experience steatorrhea which calls for a temporary suspension or reduction of the drug. Also at high dosages there are two additional complications. Since cholestryamine is a chloride form of an anion-exchange resin, the patients experience hypercholremic acidosis. Moreover, both cholestyramine and colestipol may hamper the absorption of fat-soluble vitamins. Thus, vitamin supplementation should be considered whenever appropriate. The fall in plasma prothrombin levels has also been reported. Bile acid-binding resins may be used in children with heterozygous familial hypercholesterolemia. Although the literature is limited, it is recommended to initiate therapy only after age 6. Growth curves and potential vitamin deficiency must be carefully monitored.

Nicotinic Acid
Chemistry. The structural formula of nicotinic acid is shown below. Although a vitamin, the pharmacological doses needed to cause reduction of plasma lipids are significantly above vitamin dosage.

Mechanism of Action. Administration of nicotinic acid leads to a decrease of VLDL and LDL in plasma through various mechanisms; (1) inhibition of lipolysis in adipose tissue; (2) decreased esterification of hepatic triglycerides; (3) activation of lipoprotein lipase with an attending increase of VLDL→ LDL conversion. Besides its hypotriglyceridemic action, nicotinic acid can lower the plasma levels of LDL and raise those of HDL. The mechanism for these lipoprotein changes is unclear.

Indications. Nicotinic acid is a good first-line drug for the treatment of endogenous hypertriglyceridemia. In doses of 1 g 3 times daily it reduces the plasma levels of VLDL up to 80% of pretreatment values. The effect is observed in less than 1 week and persists as long as the treatment is continued. Nicotinic acid is the drug of choice in type V hyperlipoproteinemia and it may also be used in familial and polygenic hypercholesterolemias. In such cases, plasma levels are reduced about 20% usually in 4–5 weeks. A further reduction of LDL is achieved when nicotinic acid is used in combination with bile acid-binding resins. Nicotinic acid is particularly indicated in dyslipoproteinemias associated with low plasma HDL because of its ability to raise significantly the levels of this lipoprotein.

Side Effects. Patients taking nicotinic acid for the first time experience diffuse itching and cutaneous flushing that last for about 30–40 min. These manifestations can be reduced by taking the drug at meal time, and may be prevented by 1 tablet (0.3 g) of aspirin 0.5 h before administration of the drug. Another possible way to reduce the drug reaction is to start the treatment with low doses, i.e., 100–200 mg t.i.d. and increase it progressively until the full therapeutic effect is reached. In general, after a few days of the therapy, patients no longer experience untoward effects. After prolonged administration, dyspepsia, vomiting and diarrhea can occur and also gastric acidity. Thus, caution must be used in subjects with peptic ulceration. Nicotinic acid can also cause hyperglycemia and hyperuricemia; this must be taken into account in patients with diabetes and gout. Once untoward effects ensue, reduction in drug dosage rather than discontinuance may be sufficient to avoid complication. On the other hand, the drug must be discontinued in cases where clear abnormalities of hepatic function are detected and in cases where overt hyperpigmentation and acanthocytosis nigricans become manifest. It must also be borne in mind that nicotinic acid may cause vasodilation and thus potentially contribute to the postural hypotension observed in hypertensive patients treated with hypotensive drugs. Nicotinic acid must not be used in children and pregnant women. There is no absolute contraindication for giving the drug to elderly patients, but dosage must be carefully adjusted to tolerance and body weight.

HMGCoA Reductase Inhibitors

These represent a new class of fungal metabolites with an established inhibitory action of HMGCoA reductase, the rate-controlling enzyme in cholesterol biosynthesis. The first product, compactin, was isolated in Japan by Endo in 1976 from cultures of *Penicillium citrinum*. More recently, mevinolin or monacolin K was independently isolated from cultures of *Aspergillus terreus* and *Monascus ruber* by investigators at

Merck, Sharpe & Dohme and by Endo, respectively. A review on the clinical use of mevinolin has recently appeared [54].

Chemistry. Compactin and mevinolin are both competitive inhibitors of HMGCoA reductase. Compactin differs from mevinolin by one methyl group on the hexahydronaphthalene ring. Both of these components can now be obtained by organic synthesis as can several analogues which also have the ability to affect cholesterol synthesis.

Compactin   R = H
Mevinolin   R = CH$_3$

Mechanism of Action. The primary site of action of mevinolin is the liver which is the main site of cholesterol synthesis. The drug is a competitive inhibitor of HMGCoA reductase; its action is dose dependent and reversible. About 95% of the administered drug is taken up by the liver and only a small percent reaches the systemic circulation. Although cholesterol is the precursor of steroid hormones, no effect no steroidogenesis has been reported. This may be explained by the fact that during mevinolin therapy the synthesis of mevalonate continues to occur. The drug reaches peak plasma levels in 2–3 days; about 80% is excreted in the intestine via the bile and about 10% via the urinary tract.

As mentioned above, mevinolin is a competitive inhibitor of HMGCoA reductase; its affinity for the enzyme is about 10,000 times higher than the natural substrate, HMGCoA. As a consequence, endogenous cholesterol synthesis is markedly reduced; this in turn simulates LDL receptor synthesis and receptor-mediated catabolism of LDL. Thus, LDL receptors are key for the action of mevinolin and this explains why this drug is ineffective in homozygous FH and why its action is potentiated by bile acid-binding resins. Both mevinolin and bile acid-binding resins act in a synergistic way in stimualting synthesis by the LDL receptor and cellular uptake of LDL from plasma, causing plasma cholesterol levels to drop up to 70% of pretreatment levels.

Indications. The drug that is available on the American market is mevinolin (Mevacor; Merck, Sharpe & Dohme). It is primarily indicated for patients with heterozygous familial hypercholesterolemia, either alone or in combination with bile acid-binding resins. Mevinolin could be used in other forms of dyslipoproteinemias with elevation of plasma triglycerides; however, it is not the drug of choice. Mevinolin should not be used independent of a diet and, in general, it must be considered only after medications have failed to significantly lower plasma cholesterol levels.

Side Effects. The starting dose of mevinolin is 20 mg daily at dinner time. If required, this dose may be increased 20 mg twice a day. At these levels the drug is well tolerated although in about 20% of patients mevinolin can cause a persistent increase of serum transaminase levels after 3–12 months' treatment. Patients may also develop lenticular opacities but no changes in visual activity. Diffuse myalgias, muscle tenderness and marked elevation of CPK associated with myositis have been reported. At the present, there is an uncertainty about side effects attending long-term therapy. Patients about to initiate mevinolin therapy are requested to have liver function tests and a slitlamp examination. While taking the drug, these assays must be repeated at 2- to 3-month intervals to monitor the safety of the therapy. Minor side effects have been reported such as constipation, diarrhea, dyspepsia, nausea and heartburn. However, the gastrointestinal effects are usually mild and transient and may not require discontinuance of the drug. Mevinolin should not be given to children, patients with an active liver disease or unexplained persistently elevated transaminase levels, pregnant or lactating women and those of childbearing age. The drug should also be withheld several months before pregnancy is planned. In clinical trials no significant adverse effects were seen when the administration of mevinolin was combined with that of digoxin, warfarin, beta-blockers, diuretics, anion-exchange resins and calcium blockers.

Probucol

The action of probucol in reducing plasma cholesterol levels was first reported in animal studies by Barnhart et al. [55] in 1970 and later recognized in man. The drug has a mild hypocholesterolemic effect which affects both LDL and HDL. Recently it has gained more visibility because of its antioxidant action and promotion of LDL uptake and degradation by receptor mechanism.

Chemistry. Probucol is a sulfur-containing bisphenol with strong antioxidant properties. Because of its high degree of hydrophobicity this drug and its metabolites are transported in the plasma associated with plasma lipoproteins.

$(CH_3)_3C$ — HO—⟨O⟩—S—C(CH_3)_2—S—⟨O⟩—OH — $C(CH_3)_3$, $(CH_3)_3C$, $C(CH_3)_3$

Mechanism of Action. The mechanism(s) by which probucol lowers plasma LDL is not well established. There is no evidence that probucol affects either synthesis or catabolism of LDL. However, it appears that the drug may act on the lipoprotein particle by itself. Because of its presumed protective effect on oxidation the LDL particles might be more readily degraded [56, 57]. In some cases, an unusually marked lowering of plasma HDL is seen; the mechanism is unclear but it appears related to a decreased synthesis of apo A-I [58]. Only 10% of the drug is absorbed by the intestine and the majority is excreted via the bile/fecal route. The drug slowly accumulates in the blood and adipose tissue and may be still present in the body 6 months after it is discontinued. This fact must be taken into account particularly with obese people on the drug.

Indications. Probucol should be considered in the treatment of mild hypercholesterolemias with a modest elevation of plasma LDL. Maximal effects are seen in 1–3 months. Usually the lowering of plasma cholesterol is in 10–20% range; however, in some patients, drops of total plasma cholesterol of about 50% are seen and this is usually associated with a marked drop in HDL. Thus, caution must be used in giving probucol to patients with low plasma HDL levels. These patients are ideal candidates for probucol therapy. In these cases, nicotinic acid, known to raise the plasma HDL levels, may be added to the regimen. In general, the indications for the use of this drug are poorly defined.

Side Effects. In the dosage of 2 tablets (250 mg each) twice a day the drug is in general well tolerated, particularly if taken with meals. About 10% of patients may experience gastrointestinal symptoms (diarrhea, flatulence, abdominal pain and nausea) but they are usually transient and may not require discontinuance of the drug. Reports on the occurrence of paresthesias, eosinophilia and angioneurotic edema have been reported be rather rarely. The drug is not recommended in children and during pregnancy. Among the most serious effects are the lowering of the plasma levels of HDL, particularly severe in some patients, and its persistence in the body for several months. Experimental animals with diet-induced hypercholesterolemia receiving probucol have died of heart arrhythmias;

there are no reports of this kind in man. However, it is advisable that patients on probucol therapy adhere to a low fat-low cholesterol diet and that ECG tracings are obtained at least at 6-month intervals. In the same vein, the drug should not be given to patients with recent history of myocardial damage and ventricular irritability.

Gemfibrozil

Gemfibrozil, a structural congener of clofibrate (ethyl ester of *p*-chlorophenoxyisobutyric acid), was synthesized in 1968, used in clinical trials in 1982 and it is now used in the United States primarily in hyper-triglyceridemic patients.

Chemistry. The structural formula of gemfibrozil indicates that it is a dimethylphenyl of the fibroic acid family.

$$\begin{array}{c} CH_3 \\ \\ \end{array} \quad -O(CH_2)_3C(CH_3)_2COOH$$
$$CH_3$$

Mechanism of Action. Gemfibrozil (Lopid; Parke & Davis) lowers the plasma levels of VLDL. The effect may be due to both decreased VLDL synthesis and increased degradation of VLDL [59]. The drug has also been shown to stimulte the activity of lipoprotein lipase, inhibit the lipolysis of adipose tissue triglycerides, decrease the uptake of fatty acids by the liver, effects which will all lead to a lowering of the plasma levels of VLDL. Gemfibrozil has no effect no chylomicrons of LDL whereas it can moderately raise the plasma levels of HDL by an unknown mechanism.

Indications. Gemfibrozil is primarily indicated in the treatment of endogenous hypertriglyceridemias, types III, and IV and V. In turn, it does not have a significant role in the treatment of hypercholesterolemias or hyperchylomicronemia. Gemfibrozil is rapidly absorbed by the intestinal tract and reaches peak concentrations in the plasma in 1–2 h after oral administration. Excretion occurs through the kidney as a glucuronide.

Side Effects. Although the drug is well tolerated, mild gastrointestinal symptoms (abdominal pain, diarrhea, nausea) can occur. In about 55% of the patients, mild anemia and leukopnia have been reported and also skin rash and muscle pain. The drug may cause mild hyperglycemia; thus, diabetic patients may require more insulin or oral hypoglycemic agents while on gemfibrozil. Gemfibrozil may also potentiate the effect of oral anticoagulants and, after prolonged use, favor the formation of gall-stones. Thus, patients with gallbladder disease should not receive gem-fibrozil. Once side effects appear, the regular dosage of the drug (600 mg twice daily) may be either discontinued or decreased. In the later case, full therapeutic effect may be achieved by the addition of nicotinic acid (100–500 mg 3 times daily).

Other Hypoliproteinemic Agents
Hypocholesterolemic Agents. These agents being second-line drugs are used less frequently and there is comparatively less literature on them. *Dextrothyroxine* is the optical isomer of *L*-thyroxine, the naturally occurring thyroid hormone. In dosages of 1–2 mg daily it does lower the plasma LDL levels of about 20%. However, it may increase the frequency of anginal pains, cardiac arrhythmias as well as cause symptoms of hyper-metabolism, i.e. nervousness, insomnia, etc. *Neomycin* is an antibiotic with hypocholesterolemic action when used orally. Its mechanism of ac-tion is believed to be mediated through a complex with the bile acids in the intestine thus preventing fat absorption. Neomycin given in divided doses of 0.5–2 g daily leads to a significant lowering of plasma LDL. However, it can cause important side effects such as ototoxity and nephrotoxity. The drug should be used with caution in patients with de-gree of renal insfficiency. *Beta-sitosterol* is a plant sterol structurally very similar to cholesterol (ethyl group at C-24 of the side chain). It is not absorbed by the intestine and is believed to prevent the absorption of the cholesterol entering the small intestine. The recommended dosage is 6 g daily in divided doses. Adverse effects including mild diarrhea, nausea and vomiting have been reported.

Hypotriglyceridemic Agents. *Clofibrate*, the ethyl ester of *p*-chlorophenoxyisobutyric acid, has proven useful in the treatment of en-dogenous hypertriglyceridemias, particularly types III and IV in a dose of 1 g twice daily. However, several important side effects have been recog-nized. In consequence, the drug is no longer a first-line hypolipidemic agent and is rarely used in the treatment of type III dysbetalipoproteine-mia which, in turn responds well to diet plus gemfibrozil or nicotinic acid. The main side effects are: flu-like symptoms with severe muscle cramps and tenderness, increase in the lithogenicity of the bile, cardiac arrhyth-mias and claudication. Moreover, it enhances the effect of anticoagulants. Finally, an increased incidence of tumors has been reported after long-term use.

Table V. Major secodary forms of hyperlipoproteinemia

Diagnosis	Type of hyperlipoproteinemia	Mechanism
Diabetes mellitus	high VLDL, HDL often decreased	increased production of VLDL and delayed catabolism (deficiency of LpL)
Nephrotic syndrome	high VLDL and LDL	increased production of VLDL and decreased VLDL and LDL clearance
Uremia	high VLDL	decreased VLDL catabolism
Hypothyroidism	high LDL	decreased LDL catabolism, depression of LDL receptor
Primary biliary cirrhosis	presence of lipoprotein X composed mostly of unesterified cholesteryl and phospholipids	diversion of bile into the circulation
Alcoholic hyperlipidemia	high VLDL	increased production of VLDL
Oral contraceptives	high VLDL	increased production of VLDL
Pancreatitis	high chylo, VLDL, low HDL	decreased catabolism of TG-rich particles
Autoimmune disease	high TG-rich particles	abnormal complexes between TG-rich particles and immunoglobulins
Corticosteroid therapy	high LDL	decreased LDL catabolism; depressed LDL receptor?
Antihypertensive therapy (diuretics, beta and calcium blockers)	high VLDL	decreased VLDL catabolism

Secondary Dyslipoproteinemias

It is essential to recognize that not all dyslipoproteinemias are primary in nature. The most common disorders that can lead to hyperlipoproteinemic states are: hypothyroidism, liver disorders, nephrotic syndrome, collagen disorders and dysproteinemias (table V). In these cases, improvement of the dyslipoproteinemia should follow the correction of the primary disease. Moreover, drugs like diuretics, beta or

Table VI. Primary dyslipoproteinemias

Hyperlipoproteinemias		Hypolipoproteinemias	
clear plasma	turbid plasma	low LDL	low HDL
hypercholesterolemia	hypertriglyceridemia alone or associated with hypercholesterolemia		
Familial hypercholes-terolemia, type IIa	familial hyperchylo-micronemia, type I	hypobetalipo-proteinemia	hypoalpha-lipoproteinemia
Polygenic hypercholes-terolemia, type IIa	familial hyper-triglyceridemia	abetalipo-proteinemia	Tangier disease
Hyperalphalipo-proteinemia	dysbetalipoproteinemia, type III		apo A-I/C-III deficiency
	familial combined hyper-lipidemia, type IIb, IV		
	combined elevation of chylomicron and VLDL, type V		

calcium blockers, steroids, hormones, etc., may confound the biochemical presentation of a primary hyperlipidemia. A careful medical history is always needed in these cases. In doubtful cases, family studies will help in dissecting primary from secondary components.

The careful evaluation of the patient's nutritional history should not be underestimated, particularly in cases of sporadic hypercholesterolemias where diet may be an important determinant in the genesis of the disorder. To avoid omissions, it is best to formulate a detailed questionnaire prepared through the joint input of a dietician and a physician both experienced in the treatment of lipid disorders. In the case of male patients, we find it useful to have wives help gather the dietary history. Much more difficult to evaluate are those patients who for professional reasons eat away from home. Their compliance to diets is usually difficult both because of inability to readily find suitable food and also because of the social setting attending professional activities favoring irregularities both in terms of food and alcoholic intake.

Finally, whenever a primary dyslipoproteinemia is suspected in addition to a secondary form, evaluation of all the family members should be carried out. In this context, according to our personal experience, a Lipid Clinic which operates in both adult and pediatric modes, readily facilities the differential diagnosis.

Primary Dyslipoproteinemias

A classification of primary disorders of lipoprotein metabolism is shown in table VI. Two main categories are identified: those with high levels and those with low levels of plasma lipids. In turn, the hyperlipoproteinemias can be divided into those characterized by an elevation of plasma cholesterol and cholesterol-rich lipoproteins, i.e. LDL, and those characterized by elevation of plasma triglyceride with or without overt hypercholesterolemia. In the former the plasma is clear whereas in the latter the plasma that contains large amounts of triglyceride-rich particles scatters light and appears turbid to creamy depending upon the degree of the hypertriglyceridemia. Hyperlipidemias, regardless of their nature, are in general associated with an increased incidence of CVD which may be correctable be dietary changes alone or by combining diets and drug(s). Another group of disorders classified as hypolipidemias is characterized by low levels of plasma lipids involving either low-density or high-density lipoproteins or both. Primary hypolipidemias are not associated with a high risk for CVD and may be difficult to rectify by either diets or drugs. Neurological manifestations might occur, but a causal relationship has not been established.

Hyperlipoproteinemias

Hypercholesterolemia (table VII)

Familial Hypercholesterolemia

Familial hypercholesterolemia (FH) is a disorder inherited as an autosomal-dominant trait characterized by either a partial (heterozygous form) or a total (homozygous form) absence of the LDL receptor responsible for the uptake of LDL from plasma [1, 60]. This disorder corresponds to the type IIa phenotype according to the classification of Frederickson, Levy and Lees. The frequency of the heterozygous form is about 1/500 of the general population whereas that of homozygous form is close ot 1/1,000,000. FH is expressed very early in life and can be detected even in the umbilical cord blood. In heterozygous children, plasma cholesterol levels are around 200–250 mg/dl; in adults around 400 mg/dl marked elevation of LDL and usually low HDL of the very high density type (fig. 4). Homozygous subjects may have plasma cholesterol above 400 mg/dl up to 1,000 mg/dl. Heterozygous patients by the age of 40 present with tendon xanthomas characterized by fibrous tissue and macrophages filled with cholesteryl esters (table VIII). The area most fre-

Table VII. Primary forms of primary hyperlipoproteinemia with clear plasma hypercholesterolemias

Disorder	Pattern of inheritance	Plasma lipoprotein abnormality	Proposed mechanism
Familial hypercholesterolemia heterozygous and homozygous; type IIa	autosomal dominant	high LDL; in homozygous subject HDL can be low	deficiency or absence of LDL receptor leading to delayed catabolism of LDL and IDL
Polygenic hypercholesterolemia; type IIa	unknown	high LDL	unknown; in cases where the LDL receptor is functionally normal consideration may be given to a mutation in the apo B_{100} region involved in LDL receptor binding
Familial hyperalphalipoproteinemia	autosomal dominant	high HDL, predominantly HDL_2	unknown

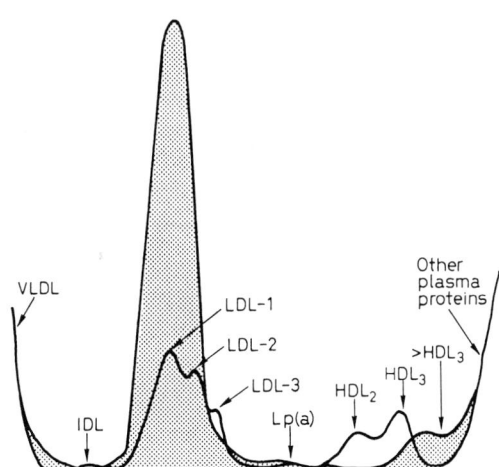

Fig. 4. 'Single-spin profile' of the plasma of a patient with homozygous FH. Note the marked elevation of LDL and the low HDL banding in a region heavier than HDL_3.

Table VIII. Types of xanthomas

Type of xanthoma	Frequent localization	Clinical appearance	Tissue pathology	Occurrence in hyperlipoproteinemias
Tendinous	extensor tendons of hands, knees, elbows and achilles tendons (sometimes difficult to distinguish from rheumatoid nodules or gouty arthritis)	deeply situated firm nodules of various sizes; move with tendons; overlying normal skin	dermal macrophages containing large amounts of free and esterified cholesterol	familial hypercholesterolemia, type IIa
Planar	several areas of the body; eyelids (xanthelasmas), palmar creases (xanthoma striatum palmare)	yellow, soft, macular or slightly elevated plaques; in palmar creases, yellow to orange linear lesions	dermal infiltrates containing both cholesterol and triglycerides	dysbetalipoproteinemia (type III); also seen in biliary cirrhosis with elevation of Lp(x)
Tuberous	extensor surfaces and in particular elbows, knees, knuckles, buttocks as well as palms	early in their evolution they appear as small, soft papules that later coalesce, enlarge, and become modular and firm with increasing fibrosis	lesions containing both cholesterol and triglyceride	dysbetalipoproteinemia (type III); also seen in biliary cirrhosis
Eruptive	extensor surfaces of the arms, legs and buttocks (xanthoma diabeticorum when associated with diabetes mellitus)	small, yellow papules with an erythematous halo around the base; may appear in crops over pressure points	triglyceride-rich lipoproteins from plasma traverse the dermal capillaries, are taken up by dermal histiocytes which then become foam cells; cells contain both cholesterol and triglycerides	lipemic serum; elevation of chylomicrons and/or VLDL; seen in type I, IV and V

quently involved is the calcaneal followed by the estensor tendons of the hands, patella and the extensor tendons of the feet. Tendon xanthomas may become inflamed. The inflammation usually lasts for 1–2 weeks during which swelling and tenderness is present. FH patients may also present with modular (tuberous) xanthomas most frequently found around the elbows, knees, ankles and hands. In homozygous subjects, particularly frequent are planar xanthomas which are patchy, yellow skin infiltrates over the trunk and the extremities. Cholesteryl ester infiltrates around the eyelids (xanthelasmas) and corneal arcus also occur in FH. Arcus is diagnostic if observed before age 40; at a later age, it may be present independent of hypercholesterolemia.

A most serious complication of FH is coronary heart disease. In heterozygous subjects, it usually appears in the third to fourth decade of life and slightly later in women. On the other hand, homozygous subjects may experience coronary heart disease in the first decade and may die before the second decade of life.

The genetic basis of FH has been identified at the level of the LDL receptor by the classic work of Brown and Goldstein [60]. From recent studies, it appears that a familial form of hypercholesterolemia might occur in subjects with normal LDL receptor function but a mutation in the apo B_{100} domain involved in LDL receptor binding [61].

The diagnosis of FH is made on the basis of persistent elevation of plasma cholesterol and LDL levels usually unaffected by diets, by the history of hypercholesterolemia in other family members and by either a personal and/or family history of atherosclerotic cardiovascular disease. When in doubt, cell culture studies should be carried out to establish the defect in LDL receptor function. However, such a test is only available in specialized centers and heterozygous subjects may be difficult to differentiate from controls. The usefulness of RFLPs using DNA isolated from lymphocytes of peripheral blood has not yet been established.

The treatment of heterozygous FH combines the use of phase 1 or 2 AHA diet (phase 3 is usually not well tolerated for long periods) and drugs. In children below age 6, diet alone is recommended; after age 6 bile-acid binding resins may be used in dose of 12 g (cholestyramine) and 15 g divided into three doses. In adults, ion-exchange resins are given with nicotinic acid (1 gram 3 times daily). This drug combination is usually effective but occasionally may require the addition of probucol (500 mg twice daily). The more recent entry in the treatment of FH is mevinolin (20 mg daily or twice daily) in combination with cholestyramine or colestid. The results obtained thus far are highly encouraging. In the case of homozygous patients, medical treatment is not rewarding. Chronic programs of plasmapheresis of LDL-pheresis may be beneficial but

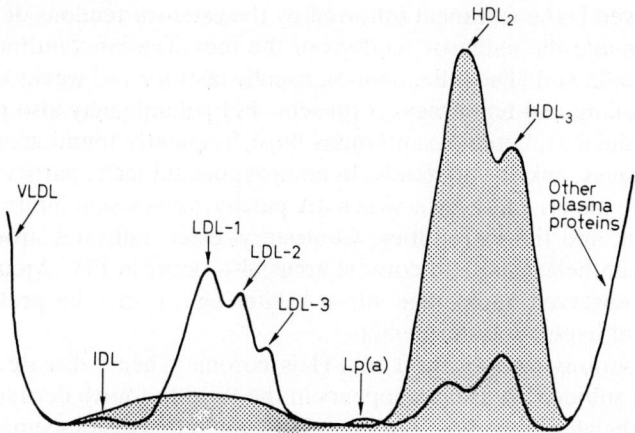

Fig. 5. 'Single-spin profile' of the plasma of a subject with hyperalphalipoproteinemia. Note the marked elevation   HDL$_2$. The subject had an HDL cholesterol near 100 mg/dl and apo A-I of 180 mg/dl. Note the low LDL and presence of Lp(a).

are demanding in time, are invasive, costly and still limited to specified centers. These patients are ideal candidates for liver transplant and there is a positive experience in this regard.

Polygenic Hypercholesterolemia

Under the IIa phenotype there are also the polygenic hypercholesterolemias [9] for which an actual mechanism of action has not been established although it is likely to represent more than a single disease entity. The age of onset is variable and the elevation of plasma cholesterol and LDL are not associated with an elevation of triglycerides. Xanthomas and xanthelasmas are usually absent and the association with CVD is not as established as in FH. From the therapeutic standpoint it is important to determine the response of these subjects to the AHA diet starting with phase 1 for 3 months and, if ideal response is not achieved, move to phase 2 or even 3 for an additional 3–6 months. The combination of cholestyramine or colestid and nicotinic acid is usually effective provided that the dietary program is maintained. Mevinolin alone or with bile acid-binding resins should also prove beneficial.

Familial Hyperalphalipoproteinemia

Familial hyperalphalipoproteinemia is a rare disorder characterized by high levels of plasma HDL above the 70–80 mg/dl level [62, 63]. In normolipidemic subjects, HDL$_3$ is the major species. In hyperalphalipoproteinemia, the predominant form is HDL_2, whereas the HDL_3 levels are in normal limits (fig. 5). This lipoprotein abnormality is believed to be transmitted as an autosomal-dominant trait with a very high degree of penetrance. Moreover, it has been associated with an increased life expectancy and a low incidence of cardiovascular disease. Unfortunately, the rare incidence of hyperalphalipoproteinemia precludes a critical assessment of its postulated health benefits, particularly in cases where plasma levels of HDL are well above the 100 mg/dl range. In fact, a recent preliminary report has appeared associating hyperalphalipoproteinemia with heart disease [64]. The biochemical or genetic basis for the hyperalphalipoproteinemia has not been determined. No treatment is indicated. Secondary causes leading to an elevation of plasma HDL levels must be ruled out, i.e. estrogen or nicotinic acid therapy, intense aerobic exercise program associated with weight loss. The diagnosis of primary hyperalphalipoproteinemia requires verification of its familial occurrence.

Hypertriglyceridemias (table IX)

Primary Hyperchylomicronemia (Type 1 Phenotype)
This is a rare disorder [63, 65] usually expressed in childhood and characterized by a marked elevation of fasting plasma triglycerides and chylomicrons, low HDL cholesterol and apo A-I. Patients have usually bouts of abdominal pains. When plasma triglycerides reach levels in the 2,000–3,000 mg/dl range, acute pancreatitis may ensue. There are two forms of primary hyperchylomicronemia: one due to lipoprotein lipase deficiency and the other to apo C-II deficiency. The two entities will be examined separately.

LpL Deficiency. This rare disorder is characterized by a deficiency of LpL which is inherited as an autosomal-recessive trait with an incidence of about 4 in 1 million persons. The plasma of these patients is milky in appearance in the fasting state; the chylomicrons are markedly high, VLDL either normal or slightly elevated and LDL and HDL cholesterol are low. The patients present with lipemia retinalis and eruptive xanthomas, cutaneous lesions containing macrophages filled with triglycerides and cholesteryl esters. Liver and spleen are enlarged due to the accumulation of fat-filled macrophages. Abdominal pain is usually present and is particularly intense during episodes of acute pancreatitis. The mechanism for the pancreatitis is still unclear but it is probably related to capillary blockade by chylomicron clumps leading to death of pancreatic

Table IX. Major forms of primary hyperlipoproteinemias with turbid plasma

Disorder	Pattern of inheritance	Plasma lipoprotein abnormality	Proposed mechanism
Lipoprotein lipase deficiency or apo C-II deficiency	monogenic; autosomal recessive	creamy plasma due to massive increased levels of chylomicrons; HDL levels are low; in apo C-II deficiency the hyperlipidemia is less severe	impaired hydrolysis of chylomicron core triglycerides secondary to deficiency of lipoprotein lipase or of its cofactor apo C-II
Dysbetalipoproteinemia; type III	monogenic?	elevation of chylomicron remnants and IDL	decreased mechanism of degradation of chylo remnants and LDL; it requires homozygosity for apo E_2 phenotype and additional factors (diabetes, obesity, etc.) for hyperlipidemia to be expressed
Familial hypertriglyceridemia, type IV or V	autosomal dominant	elevation of VLDL	unknown
Familial combined hyperlipidemia, type IIb or IV	autosomal dominant?	elevation of VLDL and LDL; variable expression in family members	unknown
Type V hyperlipoproteinemia	autosomal dominant?	elevation of chylo and VLDL; low HDL	unknown
Hepatic lipase deficiency	uncertain	elevation of VLDL	deficiency of hepatic lipase

acinar cells, release of pancreatic enzyme and autolysis. Patients injected intravenously with heparin (50–100 units/kg) exhibit very low or no LpL activity due to either a deficiency of intracellular enzyme or defective cellular release. The post-heparin HL activity is normal.

The treatment of this genetic disease is aimed at reducing the plasma triglycerides to levels below those known to cause pancreatitis (i.e. around 1,000 mg/dl or less). This is achieved by restricting fats to about 10% of the total calories using diets relatively enriched in carbohydrates. Patients may also be given medium-chain triglycerides which by being water soluble are absorbed through the portal vein and do not form chylomicrons. Drug therapy is ineffective.

Apo C-II Deficiency. The first report on a genetically determined hyperchylomicronemia due to a deficiency of apo C-II, the cofactor of lipoprotein lipase, was in 1978 by Breckenridge et al. [66] in a family of 14 affected members. Three other kindreds have been reported since. This is a relatively rare disease with an autosomal-recessive pattern of inheritance. Marked hyperchylomicronemia is present in homozygous subjects whereas heterozygotes have normal plasma lipoprotein levels. The overall clinical presentation is similar to that of LpL deficiency except that the hyperchylomicronemia is usually less severe, VLDL levels may be elevated and eruptive xanthomas are not commonly present. These findings and the absent or modest hepatosplenomegaly indicate that in apo C-II deficiency the defect in chylomicron metabolism is comparatively less severe than in LpL deficiency. the diagnosis is based on the lack of LpL activity in post-heparin plasma and by the restoration of this activity in the presence of either normal serum or pure apo C-II. The more direct documentation of the disease is the total absence of apo C-II in chylomicrons or the presence of a functionally abnormal apo C-II. The treatment of apo C-II deficiency is the same as for lipoprotein lipase deficiency.

Familial Combined Hyperlipidemia

This familial disorder was discovered by Goldstein et al. [10] when studying familial members of patients with myocardial infarction presenting with either increase VLDL or LDL or both. Even in the same individual the pattern of dyslipoproteinemia may vary from time to time. This unstable lipoprotein abnormality is rarely present in children, is preferentially expressed in middle life or later and is transmitted in a Mendelian dominant pattern.

Patients with familial combined hyperlipidemia are at a great risk for atherosclerotic CVD. This may be related to the fact that these patients have been reported to have an increased production and secretion of

VLDL apo B_{100} leading to increased levels of apo B_{100}, one of the characteristics of familial combined hyperlipidemia. Elevation of plasma levels of apo B_{100} is also present in a lipoprotein disorder referred to as 'hyperapobetalipoproteinemia'. First believed to represent an entity by itself, it is now considered to be a variant of familial combined hyperlipidemia. Overall the genetic basis of familial combined hyperlipidemia is poorly understood; it likely represents a heterogeneous disorder.

From the therapeutic standpoint, diet should be the first choice and in obese patients directed at the normalization of body weight with reduction in simple sugars and alcoholic intake. The regular intake (3–4 times/week) of fish, like salmon which is rich in w-3 fatty acids, should be encouraged. If diet alone does not achieve the desired goals, drugs should be added. Gemfibrozil and nicotinic acid alone or in combination are usually effective. Nicotinic acid is particularly indicated when low plasma HDL levels are present.

Familial Hypertriglyceridemia

In this disorder, patients have a fasting endogenous hypertriglyceridemia, which is also found in family members [63, 65]. A single elevation of plasma triglyceride levels in the proband alone is not sufficient for diagnosing this disorder. Plasma triglycerides may be elevated, either in association with other inherited disease, for instance hypoalphalipoproteinemia, or as a secondary manifestation of nongenetic abnormalities. In true familial cases, expression of the disease is usually encountered after age 20, and hypertriglyceridemia is a constant finding.

Some patients show a disease pattern similar to that described for type IV hyperlipoproteinemia; this may, however, be only a chance occurrence and not a genetic relationship. Familial hypertriglyceridemia is transmitted as an autosomal-dominant trait; the nature of the genetic defects is unknown. As in other inherited disorders, the contribution of nongenetic factors to hypertriglyceridemia may show some degree of obesity, abnormalities in insulin metabolism, hyperglycemia, and hyperuricemia. Sorting the primary from the secondary aspects of the elevation of plasma triglyceride levels may pose a problem. An increased risk of premature coronary heart disease has been reported. In a study by Hazzard et al. [11], about 38% of the first-degree relatives of probands with familial hypertriglyceridemia died of myocardial infarction, compared with 18% of the deceased relatives of normolipidemic controls.

The pathogenesis of this disorder has not been established. Abnormalities in either the production or the decreased removal of triglycerides have been postulated. In overproduction, hyperinsulism due to insulin resistance has been suggested, and in defective removal, an extrahepatic

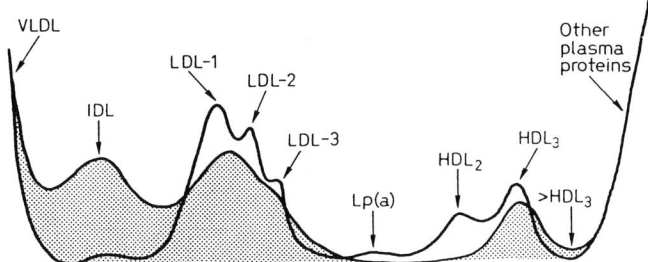

Fig. 6. 'Single-spin profile' of the plasma of a patient with dysbetalipoproteinemia (type III). Note the increase of VLDL-IDL region and the low HDL (HDL$_3$ and heavier).

mechanism may be present; no defect in lipoprotein lipase activity has been detected, however. It is possible that familial hypertriglyceridemia is the expression of two different pathogenic disorders. Should this be the case, the single gene mode of inheritance may be open to question. There is obviously a need for further studies to identify the nature of the regulatory defect in these patients. Since they may have either type IV or V pattern, the possibility that components of the elevated plasma lipid levels may be due to exogenous sources must be ruled out.

In view of the many variables, familial hypertriglyceridemia is difficult to diagnose and its genetic nature can only be established by the study of large family groups. If fasting hypertriglyceridemia is observed in the patient under study, a careful search for nongenetic causes must be undertaken.

Treatment is directed at a diet low in simple sugars either alone or combined with a low-calorie regimen. Fish oils may be considered. The drug of choice is gemfibrozil and it should be used in combination with dietary therapy.

Dysbetalipoproteinemia

Dysbetalipoproteinemia, also known as type III hyperlipoproteinemia, is a disorder characterized by the elevation of IDL (fig. 6) containing about equal amounts of cholesterol and triglycerides and migrating in β position on agarose gel electrophoresis, as contrasted to the pre-β mobility exhibited by normal VLDL [63]. Terminologies also used in the past are broad-beta or floating beta disease [67]. This disorder is associated with a high incidence of atherosclerotic CVD. About 50% of these patients present with planar xanthomas localized at the palmar creases as well as about the elbows, knees and buttocks.

Patients with dysbetalipoproteinemias have plasma cholesterol levels between 300 and 500 mg/dl and triglycerides values in the same range. Both VLDL and IDL are enriched in cholesterol in relation to triglycerides with ratios as high as 1.0. The major metabolic defect appears to be the failure by these patients to further process chylomicrons and VLDL remnants. Apo B_{48} is present in chylomicron remnants which also have an increased content in apo C-III and apo E. The latter is characterized by the presence of the apo E 2/2 genotype [68, 69] which is structurally less suited for recognition by the remnant and LDL receptors than the other apo E phenotypes [70]. However, this genetically determined structural abnormality of apo E is by itself insufficient to cause the characteristic elevation of plasma cholesterol and triglycerides. This is documented by the observation that the incidence of clinically manifested dysbetalipoproteinemia is only 1–2/10,000 whereas the apo E 2/2 genotype is 1/100 individuals. Thus, additional factors are required for the expression of the disease and they may be acquired (i.e. diet, alcohol, disease states, such as hypothyroidism, etc.) or genetic. In the latter respect, dysbetalipoproteinemia has been found to be associated with either familial hypercholesterolemia of familial combined hyperlipidemia. However, other unrecognized genetic factors may also be present. Thus, dysbetalipoproteinemia must be viewed as a multifactorial disorder in which the age factor also plays a role. this disorder is rather amenable to therapy. Diet should be aimed at restoring ideal body weight. The intake of simple sugars must be reduced, alcohol discontinued and a structured aerobic exercise program recommended; the use of a diet enriched in w-fatty acids may be encouraged. When these measures do not reach the ideal goal, drugs must be added to the diet. The most effective in this regard is nicotinic acid used alone or in combination with gemfibrozil. Mevinolin may also be used but it is not a first-choice drug.

Hepatic Triglyceride Lipase Deficiency

This rare disorder characterized by a deficiency of HL may present with a plasma lipoprotein profile similar to that of dysbetalipoproteinemia. Of the two brothers reported by Breckenridge et al. [71] in 1982, one had an elevation of both plasma cholesterol and triglycerides, eruptive xanthomas in the palmar creases, corneal arcus and coronary heart disease. The older brother also presented with a mixed hyperlipidemia but had neither xanthomas nor vascular disease. In spite of the elevation of cholesteryl ester-rich remnant particles (small VLDL and IDL) and the increased content in apo Cs and apo E, the apo E 2/2 genotype was absent. In addition, contrary to dysbetalipoproteinemia, both LDL and HDL cholesterol were in the normal ranges. Only the plasma trigly-

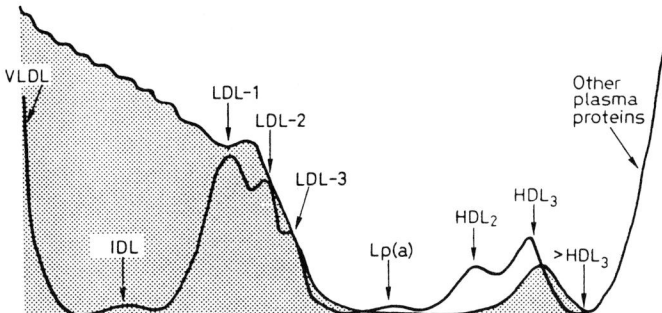

Fig. 7. 'Single-spin profile' of the plasma of a patient with type V hyperlipoproteine-mia. Note the large amount of light-scattering material in the chylo-VLDL-IDL region and low HDL.

cerides were elevated. According to Breckenridge et al. [71], the pattern characterized by the elevation of both plasma small VLDL and IDL and the relative accumulation of triglyceride-rich LDL and HDL is characteristic of hepatic lipase deficiency. This conclusion supports the concept that this enzyme is involved in the degradation of the triglycerides that accumulate in cholesteryl ester-rich particles like LDL and HDL as a consequence of the action of the cholesteryl ester/triglyceride exchange/transfer protein. The genetics of this disorder are still unknown. Since it is rare, experience on its treatment is limited. In principle, the same approach used for dysbetalipoproteinemia should be applicable.

Type V Hyperlipoproteinemia

This disorder is characterized by an elevation of both plasma chylomicrons and VLDL (fig. 7) with fasting total triglycerides reaching levels of 3,000–6,000 mg/dl or above [63, 65]. Usually this disorder is not expressed until age 30 and has a prevalence of about 2/1,000 subjects. Its pathogenesis is not clearly defined but there is no significant deficiency of either LpL, apo C-II, or HL. It is not uncommon for these patients to have diabetes, obesity, hypothyroidism and have a life-style characterized by excessive food and alcohol intake. Lipemia retinalis, eruptive xanthomas, heptatosplenomegaly and abdominal pain are usually present. Not uncommonly, patients are first seen because of acute pancreatitis, sometimes recurrent. Because patients with type V hyperlipoproteinemia are limited in number and also due to the severity of the symptoms, a vigorous hypolipidemic program is usually promptly instituted. It is difficult to establish whether this disease is associated with a high risk for atherosclerotic CVD. This assessment is also made difficult by the fact that additional independent risk factors may be present (i.e., diabetes, obesity, etc.) inclusive of low plasma levels of HDL.

The genetic basis of this disorder is unknown; it is likely to be polygenic or, if monogenic, its expression may be dependent on various exogenous factors.

The treatment calls for weight reduction, if applicable. Weight loss is better achieved by a combination of a diet and a structured exercise program. Fats must represent about 10% of the total calories and restriction of simple sugars is also recommended together with total abstinence from alcohol. The regular intake (3–4 times/week) of fish, like salmon which is rich in eicosapentanic acids, may prove helpful. In the case of acute pancreatitis, patients should be hospitalized and nutrients provided parenterally, until the hypertriglyceridemia is substantially reduced. Drugs are usually a necessary adjunct to a diet. In severe hypertriglyceridemia, nicotinic acid up to 5 g daily in three divided doses can be very effective. Drug dosage may be reduced when plasma triglyceride levels are below 1,000 mg/gl. Nicotinic acid has in addition the beneficial effect of increasing the levels of plasma HDL. Another effective drug is gemfibrozil, 600 mg twice daily, to be taken alone or in combination with nicotinic acid. Hypocholesterolemic agents such as cholestyramine, colestid or probucol are not indicated. In severe hypertriglyceridemia with an impending risk of pancreatitis we find it useful to perform up to three cycles of plasma exchange until plasma triglyceride levels return to normal values. Thereafter, the patient can be successfully treated with the combined diet and the drug combination outlined above.

Hypolipoproteinemias

Previous reviews on the subject have appeared in the literature [72, 73]. In disorders affecting *apo B-containing lipoproteins* there is a decrease (hypobetalipoproteinemia) or total absence of apo B_{100} in the plasma (abetalipoproteinemia). Patients with *hypobetalipoproteinemia* are usually asymptomatic; only a few cases have been reported presenting with neurological manifestations. However, a causal relation between these manifestations and low LDL levels has not been established. Steinberg et al. [74] described a kindred presenting with hypobetalipoproteinemia and normotriglyceridemia. Subsequent studies have established that these subjects have an abnormal species of apo B, apo B_{37} with a molecular weight of 203,000 [75] partially responsible for the low levels of plasma LDL [76].

Patients with *abetalipoproteinemia*, besides the total absence of the plasma in any form of apo B, present with fat malabsorption, abnormal red cell morphology (acanthocytosis), retinal disease (retinitis pigmentosa) and neurological symptoms (Friedreich's ataxia). If the disease is diagnosed early in life, a low-fat diet supplemented with vitamin E largely prevents all of the untoward manifestations.

Disorders expressing with *low plasma concentrations* of HDL are familial hypoalphalipoproteinemia, Tangier disease and apo A-I/C-III deficiency.

Familial hypoalphalipoproteinemia is characterized by low levels of plasma HDL cholesterol and apo A-I, normal LDL cholesterol and apo B and mild elevation of plasma triglycerides. It is transmitted in an autosomal-dominant mode and is associated with early myocardial infarction. Its prevalence is not well established but it may be more common than currently recognized. The diagnosis requires family studies. There is no specific treatment except for recommending a diet low in simple carbohydrates, low in polyunsaturated and high in monounsaturated fats together with a daily structured aerobic exercise program.

Tangier disease is a familial disorder presenting with low plasma cholesterol levels, very low HDL and a marked decrease in apo A-I having approximately equal amounts of pro-apo A-I and the mature apo A-I. Plasma triglycerides may be moderately elevated and are probably a reflection of the low plasma HDL levels. The patients have yellow-orange tonsils, splenomegaly and mild neuropathy. The mode of transmission appears to be autosomal recessive. There is no known treatment except for symptomatic measures.

Apo A-I/C-III deficiency is a genetic defect characterized by the absence of both apo A-I and apo C-III in the plasma [77] due to an abnormal rearrangement of the two genes, namely an insertion of a portion of the apo C-III gene into the apo A-I gene [78, 79]. This disorder has been described in two brothers, both of whom had severe atherosclerosis in their middle age. An autosomal-codominant mode of inheritance has been proposed. There is no specific medical treatment.

Clinical Significance of Lp(a)

In spite of the reasonable epidemiological evidence that high levels of plasma Lp(a) represent an independent risk factor of atherosclerotic CVD [32, 33] there are still many clinical uncertainties regarding the management of subjects with high level of plasma Lp(a). Of the obstacles that have prevented important advances in this area, three in particular

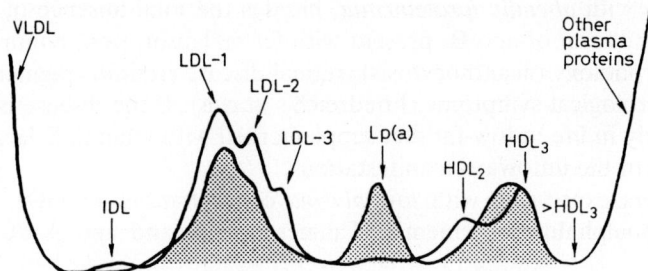

Fig. 8. 'Single-spin profile' of the plasma of a nomotriglyceridemic subject with an elevation of Lp(a).

emerge: (1) lack of standard assay for plasma Lp(a); the recent discovery of an immunological cross-reactivity between the apo(a) antigen and plasminogen [25, 26] has raised previously unperceived technical problems in this regard; (2) lack of precise criteria for defining plasma Lp(a) levels associated with a high risk for CVD; (3) limited awareness of Lp(a) among practitioners and even specialists in the area of lipid disorders. However, with the better understanding gained on the structural properties of Lp(a) and apo(a), we now expect that progress in the clinical area will also take place. In the meantime, some general guidelines apply. First, the presence of Lp(a) should be suspected in patients with a normal or borderline elevation of plasma cholesterol and a personal or familial history of CVD. Second, because of the poor response of Lp(a) patients to diets, those at a high risk for CVD may be good candidates for receiving first-choice drugs such as acid-binding resins and either probucol, nicotinic acid, or mevinolin with the realization that apo B_{100}-apo(a)-containing Lp(a) are less responsive than apo B_{00}-containing LDL. Whether an aggressive drug intervention is beneficial in reducing plasma Lp(a) levels and risk for CVD needs to be established. A typical plasma lipoprotein profile of an Lp(a)[+] subject obtained by a single-step ultracentrifugal procedure is shown in figure 8. This subject had a total plasma cholesterol around 240 mg/dl, normal triglycerides and extensive CVD as indicated by coronary bypass surgery and later angioplasty. Cases of this kind are not uncommon among Lp(a)[+] patients with severe CVD and sometimes also present with low HDL levels. In Lp(a) subjects, in general, the goal is not only to lower plasma Lp(a), but also reduce all existing risk factors, i.e. cigaret smoking, hypertension, obesity, diabetes, etc. An investigation of the first-degree relatives is also in order based on the notion that Lp(a) is transmitted in an autosomal-dominant mode [33, 34].

Overall, the development of accurate means of quantitation of plasma Lp(a) is a number-one priority and so is the establishment of the mechanism(s) whereby Lp(a) is atherogenic. In this context very important is the recent discovery that apo(a) has a striking structural similarity with plasminogen. We now need to assess the clinical significance of this observation and determine whether the plasminogen-like nature of apo(a) renders the Lp(a) particles atherogenic or, perhaps, thrombogenic. The fact that both of these proteins are polymorphic in nature should stimulate a more detailed inquiry into the genetic nature of this polymorphism. Moreover, Lp(a) is also present in the plasma of patients with hypertriglyceridemia. The characteristics of these Lp(a) particles and their potential pathogenic role is undetermined. Finally, we should consider the possibility that Lp(a) may have a physiological function. Currently there are no data in this regard.

Concluding Remarks and Perspectives

One of the major goals in the diagnosis of lipid disorders is to establish the relative contribution of genetic and environmental factors. In disorders like FH the genetic defect is at the level of the LDL receptor and, mechanistically, it is well understood. The genetic abnormality is also known in LpL and apo C-II deficiency as well as LCAT deficiency. However, this is not the case for other dyslipoproteinemias for which our understanding is comparatively more modest and polygenic and environmental factors may all contribute to the expression of the disease, for instance, type III, type V and familial combined hyperlipoproteinemia. Obviously, we need to acquire more basic knowledge before we can understand the molecular nature of each disorder. This also applies to hypolipoproteinemias. The recent structural and biological developments are now permitting us to look at apolipoproteins in more depth and in cases like apo(a) and apo E their total plasma levels must be complemented by isoform analyses. As time goes by, the knowledge of isoform patterns may also be required for other apolipoproteins inclusive of apo B. Genetic studies are being favored by the availability of specific DNA probes, and RFLP analyses have already seen application to the study of lipoprotein abnormalities [80]. However, it is too early to evaluate their overall importance and impact. The study of patients with lipid disorders has become multidisciplinary in nature and has stimulated the search for the mechanism for their association with high risk for athero-

Table X. Cholesterol: risk categories and recommended action

Total blood cholesterol level	Recommendation
Desirable level Less than 200 mg/dl of blood	recheck cholesterol level every 5 years
Borderline high 200–239 mg/dl	restrict total fats, saturated fats and cholesterol in diet and recheck plasma cholesterol level annually
Borderline high with risk factors Recommendation should be followed by men if one risk factor is present, by women if two are present, and by anyone who has had coronary heart disease	further analysis to determine level of LDL cholesterol followed by initiation of a stringent dietary program and, if necessary, by further treatment with drugs
High 240 mg/dl or higher	same as above for Borderline High with Risk Factors

sclerotic cardiovascular disease. The preventive aspect must also be considered. Primary hyperlipoproteinemias are genetically determined and transmitted. The discovery of a patient with a primary lipid disorder calls for a study of first-degree relatives and of the children and adolescent population in the family. We are living in a time where there is more awareness of the 'cholesterol problem' and practitioners are being exposed to educational materials through the joint efforts of the AHA and the National Heart, Lung and Blood Institute that have recently issued precise guidelines in terms of steps to take when plasma cholesterol levels are desirable, borderline high or high (table X). In the meantime, more knowledge is being gained on the mechanism of action of diets and new drugs with a precise mechanism of action like the HMGCoA reductase inhibitors, have entered the market. In this climate of intense fervor, it is reasonable to predict that important further gains will be made in our knowledge of lipoprotein disorders both in terms of their diagnosis and management. We should acquire the technical ability to readily identify apolipoprotein mutants and understand their role in dyslipoproteinemias and understand their potential as risk factors for vascular disease. This also applies to membrane receptors and lipolytic enzymes. The study of patients with lipid disorders should lead to a better understanding of lipoprotein metabolism in general; a key for future technical and conceptical progress in this area.

Acknowledgements

The original work in the author's laboratory that led to the development of the density gradient unltracentrifugation technique now in routine use in the study of patients with lipid disorders was supported by USPHS-HL grant 18577. Many of the ideas presented in this review have originated from many discussions with colleagues in the Lipoprotein Study Unit and the Lipid Clinic at The University of Chicago. The author wishes to thank Drs. A. Lorincz and K. Soltani of the Dermatology Division of the University of Chicago for sharing their views on the type and pathology of skin lesions attending lipid disorders. The author wishes to acknowledge the outstanding technical support provided throughout the years by Ms. Maria Garcia and Ditta Pfaffinger within the Lipoprotein Core Laboratory. The preparation of the manuscript would not have been possible without the skillful assistance of Ms. Barbara Kass and the superb support of Ms. Sherra McIntyre who put long hours in typing and editing the manuscript in a truly unselfish way.

References

1   Goldstein, J. L.; Brown, M. S.: Familial hypercholesterolemia; in Stanbury, Fredrickson, Goldstein, Brown, The metabolic bases of inherited disease, vol. 5, pp. 672–712 (McGraw-Hill, New York 1983).

2   Khachadurian, A. K.: The inheritance of essential familial hypercholesterolemia. Am. J. Med. *37:* 402–407 (1964).

3   Wilkinson, C. F.; Hand, E. A.; Fliegelman, M. T.: Essential familial hypercholesterolemia. Ann. inter. Med. *29:* 671–686 (1948).

4   Wilkison, C. F.: Essential familial hypercholesterolemia; cutaneous metabolic and hereditary aspects. Bull. N. Y. Acad. Sci. *26:* 670–685 (1950).

5   Oncley, J. L.: Plasma lipoproteins; in Folch-Pi, Bauer, Brain lipids and lipoproteins, and leucodystrophies, pp. 1–280 (Elsevier, Amsterdam 1963).

6   Lindgren, F. T.; Elliott, H. A.; Gofman, J. W.: The ultracentrifugal characterization and isolation of human blood lipids and lipoproteins with application to the study of atherosclerosis. J. phys. colloid. Chem. *55:* 80 (1951).

7   Lewis, L. A.; Olmsted, T.; Page, I. H.; Lawry, E. Y.; Mann, G. V.; Stare, F. J.; Hanig, M.; Lauffer, M. A.: Serum lipid levels in normal persons. Circulation *14:* 731–740 (1956).

8   Fredrickson, D. S.; Levy, R. I.; Lees, R. S.: Fat transport in lipoproteins – an integrated approach to mechanisms and disorders. New Engl. J. Med. *276:* 32, 94, 148, 215, 273 (1967).

9   Goldstein, J. L.; Hazzard, W. R.; Schrott, H. G.; Bierman, E. L.; Motulski, A.: Hyperlipidemia in coronary heart disease. I. Lipid levels in 500 survivors of myocardial infarction. J. clin. Invest. *52:* 1533–1543 (1973).

10 Goldstein, J. L.; Schrott, H. G.; Hazzard, W. R.; Bierman, E. L.; Motulski, A. G.: Hyperlipidemia in coronary heart disease. II. Genetic analysis of lipid levels in 176 families and delineation of a new inherited disorder, combined hyperlipidemia. J. clin. Invest. *52:* 1544–1568 (1973).

11 Hazzard, W. R.; Goldstein, J. L.; Schrott, H. G.; Motulski, A. G.; Bierman, E. L.: Hyperlipidemia in coronary heart disease. III. Evaluation of lipoprotein phenotypes of 156 genetically-defined survivors of myocardial infarction. J. clin. Invest. *52:* 1569–1577 (1973).

12 Brown, M. S.; Faust, J. R.; Goldstein, J. L.: Role of low density lipoprotein receptor in regulating the content of free and esterified cholesterol in human fibroblasts. J. clin. Invest. *52:* 1569 (1973).

13 Brown, M. A.; Goldstein, J. L.: Receptor mediated control of cholesterol metabolism. Science *191:* 150–154 (1976).

14 Scanu, A. M.; Spector, A.: Biochemistry and biology of plasma lipoproteins, pp. 1–514 (Dekker, New York 1986).

15 Gotto, A. M.: Plasma lipoproteins, pp. 1–405 (Elsevier, Amsterdam 1987).

16 Potter, A. M.; Hardman, D. A.; Schilling, J. W.; Miller, J.; Appleby, V.; Chen, G. C.; Kirsher, S. W.; McEnroe, G.; Kane, J. P.: Isolation of a cDNA clone encoding the amino-terminal region of human apolipoprotein B. Proc. natn. Acad. Sci. USA *83:* 1467–1471 (1986).

17 Knott, T. J.; Rall, S. C., Jr.; Innerarity, T. L.; Jacobson, S. F.; Urdea, M. S.; Levy-Wilson, B. J.; Powell, L. M.; Pease, R. J.; Eddy, R.; Nakai, H.; Byers, M.; Priestley, L. M.; Robertson, E.; Rall, L. B.; Betsholtz, C.; Shows, T. B.; Mahley, R. W.; Scott, J.: Human apolipoprotein B: structure of carboxyl-terminal domains, sites of gene expression, and chromosomal localization. Science *230:* 37–43 (1985).

18 Deeb, S. S.; Motulsky, A. G.; Albers, J. J.: A partial cDNA clone for human apolipoprotein B. Proc. natn. Acad, Sci. USA *82:* 4983–4986 (1985).

19 Lusis, A. J.; West, R.; Mehrabian, M.; Reuben, M. A.; Leboeuf, R. C.; Kapstein, J. S.; Johnson, D. F.; Schumaker, V. N.; Yuhasz, M. P.; Schotz, M. C.; Elovson, J.: Cloning and expression of apolipoprotein B, the major protein of low and very low density lipoproteins. Proc. natn. Acad. Sci. USA *82:* 4597–4601 (1985).

20 Carlsson, P.; Olofsson, S. O.; Bondjers, G.; Darnfors, C.; Wiklund, O.; Bursell, G.: Molecular cloning of human apolipoprotein B cDNA. Nucl. Acids Res. *113:* 8813–8826 (1985).

21 Law, S. W.; Lackner, K. J.; Hospattankar, A. V.; Anchors, J. M.; Sakaguchi, A. Y.; Naylor, S. L.; Brewer, H. B., Jr.: Human apolipoprotein B-100: Cloning, analysis of liver mRNA, and assignment of the gene to chromosome 2. Proc. natn. Acad. Sci. USA *82:* 8340–8344 (1985).

22 Yang, C. Y.; Chan, L.; Gotto, A. M., Jr.: The complete structure of human apolipoprotein B-100 and its messenger RNA; in Gotto, Plasma lipoproteins, pp. 17–94 (Elsevier, Amsterdam 1987).

23 Powell, L. M.; Wallis, S. C.; Pease, R. J.; Edwards, Y. H.; Knott, T. J.; Scott, J.: A novel form of tissue-specific RNA processing produces apolipoprotein-B$_{48}$ in intestine. Cell *50:* 831–840 (1987).

24 Chen, S.; Habib, G.; Yang, C.; Gu, Z.; Lee, B. R.; Silberman, S. R.; Cai, S.; Deslypere, J. P.; Rosseneu, M.; Gotto, A. M., Jr.; Li, W.; Chan, L.: Apolipoprotein B-48 is the product of a messenger RNA with an organ-specific in-frame stop codon. Science *238:* 363–366 (1987).

25 Eaton, D. L.; Fless, G. M.; Kohr, W. J.; McLean, J. W.; Yu, Q-I; Miller, C. G.; Lawn, R. M.; Scanu, A. M.; Partial amino acid sequence of apolipoprotein(a) shows that it is homologous to plasminogen. Proc. natn. Acad. Sci. USA *84:* 3224–3228 (1987).

26 McLean, J. W.; Tomlinson, J. E.; Kuang, W. J.; Eaton, D. L.; Chen, E. Y.; Fless, G. M.; Scanu, A. M.; Lawn, R. M.: cDNA sequence of human apolipoprotein(a) is homologous to plasminogen. Nature (in press, 1987).

27 Kirchgessner, T. G.; Svenson, K. L.; Lusis, A. J.; Schotz, M. C.: The sequence of cDNA encoding lipoprotein lipase. J. biol. Chem. *262:* 8463–8566 (1987).

28 Komaromy, M. C.; Schotz, M. C.: Cloning of rat hepatic lipase CDNA: evidence for a lipase gene family. Proc. natn. Acad. Sci. USA *84:* 1526–1530 (1987).

29 McLean, J.; Fielding, C.; Drayna, D.; Diepling, H.; Baer, B.; Kohr, W.; Henzel, W.; Lawn, R.: Cloning and expression of human lecthin-cholesterol acyltransferase cDNA. Proc. natn. Acad. Sci. USA *83:* 2335–2339 (1986).

30 Havel, R. J.: Origin, metabolic fate, and metabolic function of plasma lipoproteins; in Steinberg, Olefsky, Hypercholesterolemia and atherosclerosis: pathogenesis and prevention, pp. 117–141 (Churchill Livingstone, New York 1987).

31 Goldstein, J. L.; Brown, M. S.: Regulation of low-density lipoprotein receptors: Implications for pathogenesis and therapy of hypercholesterolemia and atherosclerosis Circulation *76:* 504–507 (1987).

32 Glomset, J. A.: The plasma lecithin-cholesterol acyl transferase reaction. J. Lipid Res. *9:* 155–167 (1968).

33 Fless, G. M.; Scanu, A. M.: Lipoprotein(a): biochemistry and biology; in Scanu, Spector, Biochemistry and biology of plasma lipoproteins, pp. 73–83 (Dekker, New York 1986).

34 Morrisett, J. D.; Guyton, J. R.; Gaubatz, J. W.; Gotto, A. M., Jr.: Lipoprotein(a): Structure, metabolism and epidemiology; in Gotto, Plasma lipoproteins, pp. 129–152 (Elsevier, Amsterdam 1987).

35 Lipid Research Clinics Program: The Lipid Research Clinics Coronary primary preventive trial results. Reduction in incidence of coronary heart disease. J. Am. med. Ass. 251–351 (1984)

36 National Institute of Consensus Conference: Lowering blood cholesterol prevents heart disease. J. Am. med. Ass. *253:* 2080–2086 (1985).

37 Treatment of Hypertriglyceridemia: NIH Consensus Conference Summary. Arteriosclerosis *4:* 296–301 (1984).

38 Nilsson, J.; Mannickarottu, V.; Edelstein, C.; Scanu, A. M.: An improved detection system applied to the study of serum lipoproteins after single-step density gradient ultracentrifugation. Analyt. Biochem. *110:* 342–348 (1981).

39 Grundy, S. M.: Dietary treatment of hyperlipoproteinemia; in Steinberg, Olefsky, Hypercholesterolemia and atherosclerosis pathogenesis and prevention, pp. 169–193 (Churchill Livingstone, New York 1987).

40 Shepherd, J.; Packard, C. J.; Grundy, S. M.; Yeshurun, D.; Gotto, A. M., Jr.; Taunton, D.: Effects of saturated and polyunsaturated fat diets on the chemical composition and metabolism of low density lipoproteins in man. J. Lipid Res. *21:* 91–99 (1980).

41 Vega, G. L.; Groszek, E.; Wolf, R.; Grundy, S. M.: Influence of polyunsaturated fats on composition of plasma lipoproteins and apolipoproteins. J. Lipid Res. *23:* 811–822 (1982).

42 Shepherd, J.; Packard, C. J.; Patsch, J. R.; Gotto, A. M.; Taunton, D. O.: Effects of dietary polyunsaturated and saturated fat on the properties of high density lipoprotein and the metabolism of apolipoprotein A-I. J. clin. Invest. *61:* 1582–1592 (1978).

43 Gofman, J. W.; DeLalla, O.; Glazier, F.; Freeman, N. K.; Lindgren, F. T.; Nichols, A. V.; Strisower, E. H.; Tamplin, A. R.: Plasma *2:* 413–484 (1954).

44 Miller, G. J.; Miller, N. E.: Plasma high density lipoprotein concentration and development of ischemic heart disease. Lancet *i:* 16–19 (1975).

45 Nestel, P. J.; Connor, W. E.; Reardon, M. F.; Connor, S.; Wong, S.; Boston, R.: Suppression of diets rich in fish oil of very low density lipoprotein production in man. J. clin. Invest. *74:* 82–89 (1984).

46 Grundy, S. M.; Abrams, J. J.: Comparison of actions of soy protein and casein on metabolism of plasma lipoproteins and cholesterol. Am. J. clin. Nutr. *38:* 245–252 (1983).

47 Melish, J.; Le, N. A.; Ginsberg, H.; Gingsberg, H.; Steinberg, D.; Brown, V.: Dissociation of apoprotein B and triglyceride production in very low density lipoproteins. Am. J. Physiol. *239:* E354–E362 (1980).

48 Knittle, J. L.; Ahrens, E. H., Jr.: Carbohydrate metabolism in two forms of hypertriglyceridemia. J. clin. Invest. *43:* 485–495 (1964).

49 American Heart Association: Diet and coronary heart disease (American Association, Dallas 1978).

50 Brown, M. S.; Goldstein, J. L.: Drugs used in the treatment of hyperlipoproteinemia; in Goodman-Gilman, Goodman, Rall, Murad, The pharmacological basis of therapeutics, pp. 827–845 (Mcmillan, New York 1985).

51 Tikkanen, M. J.; Nikkila, E. A.: Current pharmacologic treatment of elevated serum cholesterol. Circulation *76:* 529–533 (1987).

52 Hoeg, J. J.; Gregg, R. E.; Brewer, H. B.: An approach to the management of hyperlipoproteinemia. J. Am. med. Ass. *255:* 512–521 (1986).

53 Kane, J. P.; Havel, R. J.: Treatment of hypercholesterolemia. A. Rev. Med. *37:* 427–435 (1986).

54 Tobert, J. A.: New developments in lipid-lowering therapy: the role of inhibitors of hydroxymethyl-glutaryl coenzyme A reductase. Circulation *76:* 534–538 (1987).

55 Barnhart, J. W.; Sefranka, J. A.; McIntosh, D. D.: Hypocholesterolemic effect of 4-4-(isopropylidenedithio)-bis(2, 6-di-t-butylphenol) (probucol). Am. J. clin. Nutr. *23:* 1229–1233 (1970).

56 Naruszewicz, M.; Carew, T. W.; Pittman, R. C.; Witztum, J. L.; Steinberg, D.: A novel mechanism by which probucol lowers low density lipoprotein levels demonstrated in the LDL receptor-deficient rabbit. J. Lipid Res. *25:* 1206–1213 (1984).

57 Steinberg, D.: Studies on the mechanism of action of probucol. Am. J. Cardiol. *57:* 16H–21H (1986).

58 Nestel, P. J.; Billington, T.: Effects of probucol on low-density lipoprotein removal and high-density lipoprotein synthesis. Atherosclerosis *38:* 203–209 (1981).

59 Kesaniemi, Y. A.; Grundy, S. M.: Influence of gemfibrozil and clofibrate on the metabolism of cholesterol and plasma triglycerides in man. J. Am. med. Ass. *251:* 2241 (1984).

60 Brown, M. S.; Goldstein, J. L.: Familial hypercholesterolemia, a genetic defect of the low-density lipoprotein receptor New Engl. J. Med. *294:* 1386–1390 (1976).

61 Innerarity, T. L.; Weisgraber, K. H.; Arnold, K. S.; Mahley, R. W.; Krauss, R. M.; Vega, G. L.; Grundy, S. M.: Familial defective apolipoprotein B_{100}: low density lipoprotein with abnormal receptor binding. Proc. natn. Acad. Sci. USA *84:* 6919–6923 (1987).

62 Glueck, C. J.; Gartside, P.; Fallat, R. W.; Steiner, P. M.: Longevity syndromes: familial hypobeta- and hyperalphalipoproteinemia. J. Lab. clin. Med. *88:* 941 (1976).

63 Brown, V. W.; Gingsberg, H.: Classification and diagnosis of hyperlipidemias; in Steinberg, Olefsky, Hypercholesterolemia and atherosclerosis: pathogenesis and prevention, pp. 143–168 (Churchill Livingstone, New York 1987).

64 Yamashita, S.; Matsuzawa, Y.; Kubo, M.; Tarui, S.: Coronary heart disease and corneal opacification in hyperalphalipo-proteinemia. Circulation *76:* suppl., p. 120 (1987).

65 Fredrickson, D. S.; Goldstein, J. L.; Brown, M. S.: The familial hyperlipoproteinemias; in Stanbury, Wyngaarden, Fredrickson, The metabolic basis of inherited disease; 4th ed., pp. 604–655 (McGraw-Hill, New York 1978).

66 Breckenridge, W. C.; Little, J. A.; Steiner, G., et al.: Hypertriglyceridemia associated with deficiency of apolipoprotein C-II. New Engl. J. Med. *298:* 1265–1273 (1978).

67   Fredrickson, D. S.; Morganroth, J.; Levy, R. I.: Type III hyperlipoproteinemia: An analysis of two contemporary definitions. Am. J. intern. Med. *82:* 150 (1975).
68   Utermann, G.; Pruin, N.; Steinmetz, A.: Polymorphism of apolipoprotein E-III. Effect of a single polymorphic gene locus on plasma lipid levels in man. Clin. Genet. *15:* 63–72 (1979).
69   Zannis, V. I.; Breslow, J. L.: Characterization of a unique human apolipoprotein E variant associated with type III hyperlipoproteinemia. J. biol. Chem. *255:* 1759–1762 (1980).
70   Rall, S. C., Jr.; Weisgraber, K. H.; Innerarity, T. L.; Mahley, R. W.: Structural basis for receptor binding heterogeneity of apolipoprotein E from type III hyperlipoproteinemic subjects. Proc. natn. Acad. Sci. USA *79:* 4696–4700 (1982).
71   Breckenridge, W. C.; Little, J. A.; Alaupovic, P.; Wang, C. S.; Kuksis, A.; Kakis, G.; Lindgren, F.; Gardiner, G.: Lipoprotein abnormalities associated with a familial deficiency of hepatic lipase. Atherosclerosis *45:* 161–179 (1982).
72   Herbert, P. N.; Gotto, A. M.; Fredrickson, D. S.: Familial lipoprotein deficiency; in Stanbury, Wyngaarden, Fredrickson, The metabolic basis of inherited disease, vol. 4, pp. 544–588 (McGraw-Hill, New York 1978).
73   Scanu, A. M.; Cabana, V.; Spector, A. A.: Lipoprotein disorders: defects in apolipoproteins, enzymes and receptors; in Scanu, Spector, Biochemistry and biology of plasma lipoproteins, pp 453–474 (Dekker, New York 1986).
74   Steinberg, D.; Grundy, S. M.; Mok, H. Y. I.; Turner, J. D.; Weinstein, D. B.; Brown, V. W.; Albers, J. J.: Metabolic studies in an unusual case of asymptomatic familial hypobetalipoproteinemia with hypoalphalipoproteinemia and fasting chylomicronemia. *64:* 292–301 (1979).
75   Young, S. G.; Berties, S. J.; Curtiss, L. K.; Witztum, J. L.: Characterization of an abnormal species of apolipoprotein B, apolipoprotein B-37, associated with familial hypobetalipoproteinemia. J. clin. Invest. *79:* 1831–1841 (1987).
76   Young, S. G.; Berties, S. J.; Curtiss, L. K.; Dubois, B. W.; Witztum, J. L.: Genetic analysis of a kindred with familial hypobetalipoproteinemia. Evidence for two separate gene defects: one associated with an abnormal lipoprotein species, apolipoprotein B-37, and a second associated with low plasma concentration of apolipoprotein B-100. J. clin. Invest. *79:* 1842–1851 (1987).
77   Norum, R. A.; Lakies, J. B.; Goldstein, S.; Angel, A.; Goldberg, R. B.; Black, W. D.; Noffee, D. K.; Dolphin, P. J.; Edelglass, J.; Borogard, D. D.; Alaupovic, P. N.: New Engl. J. Med. 1513–1517 (1982).
78   Karathanasis, S. K.; Zannis, V. I.; Norum, A.; Breslow, J. L.: Nature *301:* 718–720 (1983).
79   Karathanasis, S. K.; Zannis, V. I.; Breslow, J. L.: A DNA insertion in the apolipoprotein A-I gene of patients with premature atherosclerosis. Nature *305:* 823–825 (1983).
80   Breslow, J. L.: Lipoprotein genetics and molecular biology; in Gotto, Plasma lipoproteins, pp. 359–397 (Elsevier, Amsterdam 1987).

Dr. Angelo M. Scanu, Department of Medicine, 5841 South Maryland Avenue, Box 231, Chicago, IL 60637 (USA)

Lusis A J, Sparkes S R (eds): Genetic Factors in Atherosclerosis: Approaches and
Model Systems. Monogr Hum Genet. Basel, Karger, 1989, vol 12, pp 50–78

Statistical Approaches to Identifying Major Locus Effects on Disease Susceptibility

Jean W. MacCluer[1]

Department of Genetics, Southwest Foundation for Biomedical Research,
San Antonio, Tex., USA

Introduction

Identification of single gene defects has led to striking progress in
understanding the pathways in normal and abnormal lipoprotein metabolism and the mechanisms of development of coronary heart disease.
Typically, the first step in establishing the monogenic basis of a disorder is
a statistical analysis, demonstrating that the pattern of occurrence of the
disease within families is consistent with the rules of simple Mendelian
inheritance.

A well-known example is provided by familial hypercholesterolemia
(FH). Family studies done more than 20 years ago suggested a monogenic
mode of inheritance for FH [Khachadurian, 1964]. Further evidence was
provided by analyses of additional families [Nevin and Slack, 1968; Goldstein et al., 1973; Kwiterovich et al., 1974]. Linkage of the FH gene to the
locus determining the third component of complement (C3) was shown
by Berg and Heiberg [1976] and by Elston et al. [1976]. Subsequently, the
work of Brown, Goldstein and their colleagues [reviewed in Brown and
Goldstein, 1986] has shown that FH is a heterogeneous group of disorders, and that the FH gene is actually a series of abnormal genes affecting the LDL receptor. The C3 and LDL receptor loci have been mapped
to chromosome 19 [Whitehead et al., 1982; Donald et al., 1984].

Other single gene defects that are associated with coronary heart disease have been reviewed by Zannis and Breslow [1985] and by Lee and
Burns [1986]. These include mutations in apolipoprotein structural genes,

[1]Supported by NIH grants PO1-HL28972 and RO1-GM31575. I am grateful to
Drs. Bennett Dyke, Candace M. Kammerer, James E. Hixson, John L. VandeBerg, John
Blangero, R. Mark Sharp, and Henry C. McGill, Jr. for helpful criticisms.

deficiencies of apolipoproteins, deficiencies of enzymes involved in lipoprotein catabolism, defects in HDL-C synthesis or catabolism, and overproduction or decreased clearance of VLDL triglyceride. For many of these defective genes, the metabolic and molecular studies that clarified the nature of the defect were preceded by statistical genetic analysis of disease patterns in families.

When a disorder is determined by alleles at a single autosomal or X-linked locus, the likely mode of inheritance frequently can be inferred by statistical analysis of family data. As long as the rules by which families are chosen for analysis are clearly specified so that corrections for biases in ascertainment can be made, comparison of observed segregation ratios with those expected under various single gene models is relatively straightforward. However, for diseases like coronary heart disease that do not have a simple genetic basis, statistical analysis is more problematic. Susceptibility to such diseases is influenced by complex interactions of genetic and environmental factors. The phenotypic data available from family members frequently include not only disease state, but also values for a variety of quantitative variables, such as serum and lipoprotein cholesterol concentrations. Only recently have methods been developed for detecting the contribution of single genes to diseases of complex etiology or to variation in quantitative risk factors. This extension of the analysis of familial patterns of disease has been one of the most important recent developments in genetic epidemiology.

The aim of this presentation is to describe statistical methods that are currently available for detecting the contribution of single genes to traits determined by both genetic and environmental factors. These traits include (1) common diseases such as coronary heart disease, and (2) quantitative risk factors such as serum concentrations of lipoproteins and apolipoproteins. Examples of applications of these methods are cited, although no attempt is made to present an exhaustive review of the literature.

The Major Locus Concept

A single genetic locus that has a detectable effect on disease susceptibility or on the level of a quantitative risk factor is termed a *major locus* (or *major gene*).

For quantitative traits, the effect of a major locus may be represented schematically as shown in figure 1. In this example, the major locus has two codominant alleles, A_1 and A_2, with frequencies p and q, respectively. Each of the three major locus genotypes, A_1A_1, A_1A_2, and A_2A_2, has a

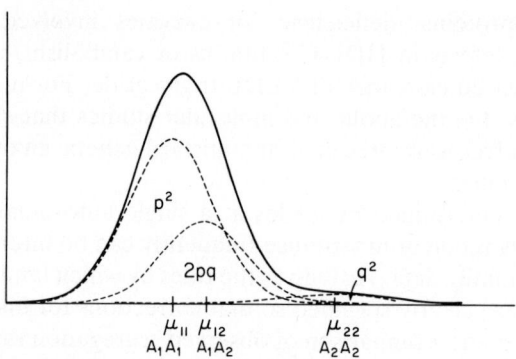

Fig. 1. Major locus model for a quantitative trait. Genotypes A_1A_1, A_1A_2, and A_2A_2 have phenotypic means μ_{11}, μ_{12}, and μ_{22}, and frequencies p^2, $2pq$, and q^2. The three dashed curves represent the distributions for each major locus genotype, and the solid curve is the sum of these three distributions.

distinct distribution of trait levels (phenotypic values). The mean level is lowest for individuals of genotype A_1A_1 (mean = μ_{11}) and highest for A_2A_2 individuals (mean = μ_{22}). For each major locus genotype, variation around the mean is determined by environmental factors and by multiple genes with individually small effects (polygenic effects). For simplicity, it is often assumed that the three major locus genotypes are in Hardy-Weinberg proportions (p^2, $2pq$, and q^2), that the environmental determinants of the trait are random (that is, they are not correlated within families and are not a function of genotype), and that the effect of polygenic factors is not a function of major locus genotype.

The effect of a major locus on a quantitative phenotype is measured in terms of its contribution to the population variance of the trait. In general, highly polymorphic loci have a larger effect on the variance than do loci with only rare variant alleles. Likewise, loci for which there is wide separation of phenotypic means contribute more than do those with little difference between means. There is no general agreement on the magnitude of the effect on phenotypic variance that is necessary to define a locus as 'major'.

For diseases with multiple interacting causes, it is useful to think of the chance of developing the disease, or *disease liability*, as a quantitative trait. This concept, which was first described by Falconer [1965], assumes that disease liability is a 'graded attribute' determined by genetic and environmental factors. Individuals are assumed to be affected if their liability falls above some threshold value. The major locus model described above for quantitative traits can thus be extended to discrete traits

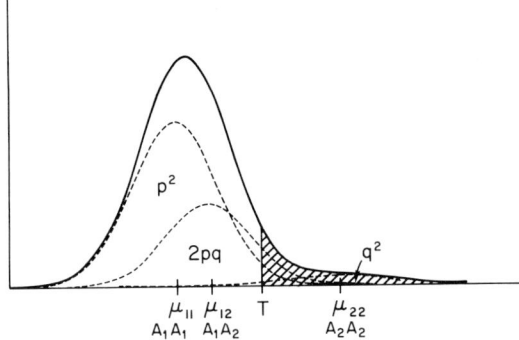

Fig. 2. Major locus model for a dichotomous trait, showing the distribution of disease liability. Individuals whose liability is greater than the threshold T (in the shaded portion of the distribution) are affected. Other symbols are as defined in figure 1.

such as diseases, for which individuals are categorized as either affected or unaffected. A major locus model for such discrete traits is represented schematically in figure 2. In this model, there is a major locus contributing to disease liability. All individuals whose liabilities fall above the threshold value T are affected. Individuals of major locus genotype A_2A_2 are most likely to be affected, and A_1A_1 individuals are least likely. The threshold model of disease liability may be extended to include two or more degrees of severity (and two or more thresholds), as is the case for familial hypercholesterolemia.

Familial Patterns of Single Gene Defects

Methods of statistical analysis developed for complex diseases are best understood by comparison with the basic methods used for suspected single gene defects. The primary criterion for determining that a defect is monogenic is a familial pattern of disease that adheres to Mendel's law of segregation. Thus, for a rare, fully penetrant dominant disorder, most affected individuals will have one affected and one unaffected parent. On average, half of their siblings will be affected and, if they have an unaffected spouse, half of their offspring also will be affected. For autosomal dominant disorders, males and females are equally likely to be affected. X-linked dominants are distinguished by their characteristic familial distribution of affected males and females. For example, affected males married to unaffected females have unaffected sons and affected daughters. For matings that are expected to produce both affected and unaffected offspring, the fit of observed data to expectations for autosomal and X-linked dominants can be evaluated by chi-square statistics.

Departures from Mendelian expectations will occur if the dominant gene exhibits reduced penetrance if affected individuals have reduced fertility. In the latter situation, a substantial proportion of affected individuals may have received a new mutation and will therefore have fewer affected relatives than predicted by Mendel's law of segregation. Departure from Mendelian ratios also will result if families are preferentially included in the study population on the basis of the number or distribution of affected family members.

Calculation of expectations for rare recessive diseases is more complicated than for rare dominants. Most individuals affected with autosomal recessive disorders have two normal (heterozygous) parents. However, in a collection of families ascertained through an affected individual, the ratio of affected to unaffected siblings will not follow Mendelian expectation (3:1) because some proportion of heterozygous couples will have produced no affected offspring and therefore will not be included in the study. Likewise, for rare X-linked recessives, a fraction of the matings of unaffected males and heterozygous females will, by chance, produce no affected sons and will escape attention. Statistical methods have been devised to correct for the failure to detect a portion of at-risk matings.

The distribution and numbers of affected family members are influenced by the way in which families are selected for study, to an even greater extent for recessive diseases than for dominant disorders. The method of statistical analysis must therefore be chosen accordingly. If a population survey is done, then all affected individuals in a population may be independently ascertained, whether or not they have affected relatives. More commonly, however, only a fraction of affected individuals come to the attention of the investigator. In this case, the choice of method of statistical analysis depends upon whether families with multiple affected individuals may be ascertained more than once. Related problems of analysis and interpretation arise if large families with multiple affected individuals are preferentially chosen for study. Even more serious is failure to document the method by which families are selected and to identify the proband(s) in each family. Without this information, corrections for ascertainment bias are impossible.

Detecting Major Locus Contributions to Complex Diseases

For complex diseases, as for rare single gene defects, statistical analysis of family data can provide clues to the existence of a basic metabolic defect. The demonstration that a familial pattern of disease is consistent with Mendelian segregation ratios is considerably more difficult for a disease with a complex etiology than for a simple autosomal dominant or recessive disorder. Susceptibility to a complex disease may be determined by numerous genetic loci, most of which have only a minor effect. There may be one or a few loci which, independently or in combination, have a major impact on susceptibility. These may interact with the minor loci and with environmental factors. Many complex diseases, such as coronary heart disease, are relatively common, and may exhibit etiologic heterogeneity. Thus, one genetic locus may have a major impact on susceptibility in some families and another locus, or an environmental variable, may play a major role in other families.

Because of these complications, statistical analyses of familial patterns of such diseases may sometimes yield ambiguous results. Therefore, the combined results of several different approaches often are used to confirm the existence of a major locus effect. Agreement among results of more than one type of analysis strengthens evidence that a major locus contributes to susceptibility to a complex disease. Among the statistical methods that are used to detect the contribution of major loci to complex diseases are (1) association analysis, in which population associations between disease status and genotype at specific loci are examined; (2) linkage analysis, in which cosegregation of the disease and specific genetic marker loci is analyzed, and (3) segregation analysis, in which the fit of family data to Mendelian segregation ratios is tested. These methods are briefly described below.

Association Studies

The demonstration that susceptibility to a complex disease is associated with genotype at a specific polymorphic locus provides preliminary evidence that the disease may be mediated by a major gene. This major gene may be either the polymorphic locus itself, or a closely linked locus that is in linkage disequilibrium with it. The loci most frequently utilized in association studies are candidate genes, that is, genes which may play a role in the disease process. Thus, for coronary heart disease, association analysis might utilize loci determining the structures of apolipoproteins, receptor loci such as the LDL receptor, or enzymes involved in lipoprotein metabolism, such as LCAT (lecithin cholesterol ester transferase), CETP (cholesterol ester transfer protein), LPL (lipoprotein lipase), and HMG CoA reductase (3-hydroxy-3-methylglutaryl coenzyme A reductase). The advantage of the candidate gene approach is that the demonstration of association between a disease and such a gene not only provides evidence for a major susceptibility locus, but also suggests a target for metabolic studies. If the chromosomal location of the candidate gene is known, the investigator also can focus on other genes in the same region. A review of the candidate gene approach is given by Lusis [1988].

As an alternative to the candidate gene approach, associations between disease susceptibility and random polymorphic marker loci (for example, loci determining serum proteins, red blood cell antigens, or DNA polymorphisms) may be examined. If disease susceptibility is influenced by a major locus that is very closely linked to a such a polymorphic marker, then alleles at the disease locus and the marker locus may be in linkage disequilibrium. As a result, specific marker alleles may exhibit increased frequencies among affected individuals.

The usual approach in an association study is to compare the distribution of genotypes, for candidate genes or for an array of marker loci, in affected individuals (cases) and in a carefully selected set of controls. Any locus that shows significant differences in allele frequency between cases and controls provides evidence for a major disease susceptibility locus. This susceptibility locus may be the marker locus or candidate gene itself, or a linked locus. A major advantage of association analysis is that it utilizes population data rather than data on affected individuals and their relatives. Thus, association studies often can be done in conjunction with other epidemiological studies, with little additional effort or expense. However, if controls are not chosen appropriately, association analyses can be misleading. For example, if controls have a different ethnic background than cases, and if marker allele frequencies differ between ethnic groups (as has often been found), then spurious associations may be revealed that are totally unrelated to a disease susceptibility gene. Problems of interpretation also may arise if associations are found in some population groups but not in others. Despite these difficulties, association analysis is a reasonable first step in the search for major disease susceptibility loci.

Numerous associations have been found between coronary heart disease or dyslipoproteinemias, and polymorphisms of apolipoprotein genes. For example, Rees et al. [1983] reported an association between hypertriglyceridemia and an apo AI gene RFLP in a British population. Ferns and Galton [1986] found that one haplotype in the apo AI-CIII-AIV gene cluster was significantly more frequent in British survivors of myocardial infarction than in controls. Ordovas et al. [1986], in a study of Caucasians in the Northeastern US, found that allele frequencies for an RFLP near the apo AI gene were significantly different between controls and patients with angiographically defined coronary artery disease,

and also between controls and patients with familial hypoalphalipoproteinemia. Associations between myocardial infarction and apo B gene
polymorphisms [Hegele et al., 1986] and between apo E polymorphisms
and dyslipoproteinemias [Leren et al., 1985; Pagnan et al., 1987] also
have been reported.

Linkage Analysis

One of the most convincing types of statistical evidence for the contribution of a major locus to disease susceptibility is the demonstration
that the disease cosegregates in families with alleles at a marker locus. As
noted above, such evidence was provided for familial hypercholesterolemia, which was shown to be linked to the locus for the third component of
complement, a locus subsequently mapped to chromosome 19.

The basic approach in linkage analysis is to calculate the likelihood of
obtaining the observed parent-to-offspring transmission pattern of a disease and a marker locus in a set of families. This likelihood is a function
of the recombination fraction between the diseases locus and the marker
locus. When two loci are closely linked on the same chromosome, they
are usually transmitted together from parent to offspring. In some proportion of individuals, recombination occurs between the two loci and
the resulting chromosome contains the paternally derived allele at one
locus and the maternally derived allele at the other.

The probability of recombination, that is, the recombination fraction,
θ, is a function of the distance between the two loci. The recombination
fraction ranges from 0.0 for loci so closely linked that recombination is
never observed, to 0.5 for loci on different chromosomes or very widely
separated on the same chromosome. In linkage analysis, the likelihood of
the observed transmission pattern in a set of families is calculated for
various values of θ. The odds of linkage for a given value of θ, say, $\theta = 0.1$, are computed by forming the ratio of the likelihood for $\theta = 0.1$ to the
likelihood for $\theta = 0.5$ (no linkage). The logarithm of this ratio, called the
log odds or lod score [Morton, 1955], is conventionally used as an indication of the strength of evidence for linkage. Traditionally, a lod score of
3.0 or greater (1,000 to 1 odds in favor of linkage) has been used as the
criterion to establish linkage, and a lod score of -2.0 or less (100 to 1
odds against linkage), to reject linkage. With the availability of data on
multiple closely linked markers in the same linkage group, lod scores
much larger than 3.0 often are obtained.

An advantage of the lod score method of linkage analysis is that results may be combined across families. The test is sequential; lod scores

may be computed separately for families from different studies, and the collection of additional families may be continued until the lod score falls outside the range −2.0 to +3.0. Although lod score calculations can be arduous for large pedigrees, programs are available to do these computations. The most widely used linkage analysis program is LIPED [Ott, 1974].

Linkage analysis is more difficult for diseases that are not simply inherited, such as coronary heart disease, than for single gene defects. For example, if disease susceptibility is determined by epistatic interaction between a major susceptibility locus and an unlinked marker locus, then linkage analysis may erroneously indicate loose linkage between the susceptibility locus and the marker [Clerget-Darpoux and Bonaïti-Pellié, 1980; Hodge and Spence, 1981]. Similarly, if a major disease susceptibility allele is segregating in some families but not in others, or if there are two different susceptibility loci, then the results of linkage analysis can be misleading. Methods have been proposed for analyzing linkage in the presence of epistatic interaction [Ott and Falk, 1982], and for detecting linkage heterogeneity [Ott, 1983; Risch, 1988]. Some of these methods have been evaluated in the Genetic Analysis Workshops [see, for example, MacCluer et al., 1984, 1985].

A potentially more serious problem is that the calculation of lod scores requires that assumptions be made about frequencies and dominance relationships of alleles at the major disease susceptibility locus, as well as probabilities that individuals of each major locus genotype are affected (that is, penetrances). Because the purpose of linkage analysis is to obtain evidence for a major locus that has not yet been identified, such information is not generally available. Therefore, linkage analysis is sometimes preceded by segregation analysis, which yields estimates of the required parameters. Alternatively, linkage analysis can be done under a range of values for allele frequencies and other parameters. The effect on linkage analysis of incorrect assumptions about parameter values has been discussed by Clerget-Darpoux et al. [1986].

When there are no clues to the location of a putative major locus determining a complex disease, investigators may follow either of the two approaches utilized in association analysis: (1) they may concentrate upon candidate genes, or (2) they may select for linkage analysis a random group of marker loci, preferably highly polymorphic and distributed among as many chromosomes as possible. Such loci may include restriction fragment length polymorphisms (RFLPs), and loci determining red blood cell and serum proteins and enzymes, blood group systems, and lymphocyte antigens. With the advent of recombinant DNA technology, the number of markers available for linkage analysis is increasing rapidly.

Considerations of cost of obtaining data dictate that the markers used in a linkage analysis be selected with care. The strategy of constructing linkage maps using random RFLPs is outlined by Botstein et al. [1980].

Multipoint Linkage Analysis

The availability of data for hundreds of polymorphic marker loci has created new opportunities for linkage analysis as well as new problems. Linkage groups containing several polymorphic marker loci can be highly informative for mapping disease loci. To place a disease locus within a known linkage map, methods such as the one developed by Lathrop et al. [1984, 1986], and implemented in the computer program LINKAGE, may be used. A major challenge has been to determine the order of multiple loci in the same linkage group. Several methods for multipoint mapping have been proposed which utilize pairwise recombination fractions for all loci in the linkage group. These methods include a multidimensional scaling approach [Lalouel, 1977], a seriation technique similar to a method used in geology [Buetow and Chakravarti, 1987a, b], and an approach based upon the product of pairwise recombination fractions [Wilson, in press]. The power of these methods to determine locus order has been investigated by Kammerer and MacCluer [in press].

Another approach to multipoint mapping is based upon estimates of the magnitude of linkage disequilibrium, which is expected to be inversely related to the distance between loci. This method, which has been used by Chakravarti et al. [1986] and others, does not require family data and may have advantages for ordering loci so closely linked that recombination is seldom (or never) observed in family studies. However, as pointed out by Weir and Hill [1986], Hedrick [1987] and others, unrealistically large sample sizes are needed to obtain precise estimates of the magnitude of disequilibrium for closely linked loci. This point is illustrated for the apo AI-CIII gene complex by Thompson et al. [1988], who reported two RFLPs in disequilibrium, while a third RFLP, located between these two, was in equilibrium with each of them.

Linkage Analysis Using Affecteds Only

Traditional methods of linkage analysis utilize data on both affected and unaffected family members. Such data are often difficult to collect, and may be unavailable for some unaffected relatives. Therefore, considerable effort is being devoted to the development of methods of linkage analysis that utilize data on affected individuals only, or on specified sets of relatives (for example, sib pairs, whether affected or not). A method that utilizes data from affected individuals in inbred pedigrees was proposed by Smith [1953] and later extended by Lander and Botstein

[1987]. With this method, linkage groups that may contain recessive disease susceptibility loci are identified by searching for chromosomal regions with many loci that are homozygous and therefore likely to be identical by descent (IBD). The major limitations of this approach are that it is appropriate only for recessive susceptibility genes and that it requires data from inbred affected individuals, who may be difficult to find in most populations.

Sib pair linkage tests [for example, Blackwelder and Elston, 1985] are based upon the expectation that sib pairs who are concordant for the disease of interest will be more alike for a linked marker locus than are sib pairs discordant for the disease. The similarity of sib pairs for the marker locus is measured by the estimated number of marker alleles that are identical by descent in the sibs. Statistical tests for linkage involve comparison of the observed IBD distributions in concordant and discordant sib pairs. An advantage of sib pair methods is that they require no assumptions about the mode of inheritance of putative susceptibility genes. However, because these methods do not yield estimates of recombination fractions, they are best used as screening methods. Sib pair methods have been extensively applied to HLA-associated diseases, such as insulin-dependent diabetes mellitus and multiple sclerosis [for example, Suarez and VanEerdewegh, 1984]. With the increasing availability of data on polymorphic marker loci, these methods now are appropriate for other complex diseases such as atherosclerosis.

More recently, methods utilizing data on other affected relative pairs, such as uncle-niece and first cousins, have been proposed. The methods differ in their approaches and in the types of data for which they are best suited. One such method [Cantor and Rotter, submitted] involves estimation of the number of marker alleles identical by descent and, in its present form, requires genotypic data for the affected relative pair and all intervening ancestors. Another method, developed by Weeks and Lange [1988], uses identity by state relations rather than identity by descent, and requires information only for the affected pedigree members. An advantage of the affected relative pairs methods of linkage analysis is that they require data on fewer individuals. The power of these methods for different types of complex diseases in currently under investigation by participants in Genetic Analysis Workshop 6.

Segregation Analysis

A major advance in the genetic analysis of disease patterns in families began with Elston and Stewart's [1971] description of a general model for detecting the contribution of genetic factors to complex traits by analysis of pedigree data. The statistical approaches described in that paper led to the development of methods of statistical analysis that are not restricted to nuclear families, and that may be applied to complex diseases or to quantitative traits. Segregation anaylsis of such traits is sometimes referred to as *complex segregation analysis*, a term introduced by Morton et al. [1971]. These methods have been implemented in several computer packages, including POINTER [Morton and MacLean, 1974], PAP [Hasstedt and Cartwright, 1979], COMBIN [MacLean et al., 1983] and SAGE [Elston et al., 1986].

In segregation analysis of family data for a complex disease, one calculates the likelihood that the familial pattern of disease would be observed under alternative hypotheses about genetic and environmental contributions to disease susceptibility. A simple hypothesis is that variation in disease liability is determined solely by random environmetal factors. The population distribution of liability for this 'sporadic' model can be characterized by a single parameter, namely, the proportion of the population who are affected. A more complex hypothesis is that variation in disease liability is determined by a major locus with two alleles, as well as by environmental factors. For this 'major locus' model, there are four parameters: the proportion of individuals of each major locus genotype who are affected (in other words, the penetrance for each major locus genotype), and the frequency of the 'disease' allele.

Evidence for a major locus contribution to disease liability can be obtained with a likelihood approach, as implemented in any one of the computer packages listed above. For example, the parameter values that maximize the likelihood of observing the pedigree data can be determined for the sporadic and major locus models, and the likelihoods for each of these two models can be compared. The statistical significance of the difference between these two models is evaluated by determining the natural logarithms of each of the likelihoods, calculating the difference between the Zn likelihoods, and multiplying by -2. This statistic is approximately chi-square distributed, with degrees of freedom equal to the difference in the numbers of parameters estimated in the two models. In this example, the number of degrees of freedom is equal to 3.

In addition to the sporadic model, with only environmental determination of the trait, and the major locus model, which includes a single codominant major locus and environmental determinants, other models also can be evaluated. Examples include a 'polygenic' model (with polygenic and environmental determinants but no major locus), a dominant or recessive major locus model (with the mean liability for the heterozygote constrained to equal the mean for one of the homozygotes), or a

'mixed' model (with major locus, polygenic, and environmental determination). The various approaches to segregation analysis that are used
in the computer packages described above differ in the types of models
which may be evaluated, in the parameters used to define these models,
in the types of family structures to which they may be applied, and in the
statistical and numerical methods used.

The use of the chi-square statistic to compare two models is restricted
to 'nested' models, that is, models for which one is a reduced version of
the other. For example, the dominant and codominant major locus models are similar except that the dominant model has a constraint not present
in the codominant: the mean phenotypic value for the heterozygote must
equal the mean for one of the homozygotes. The dominant model is thus
a reduced version of the codominant, and the significance of the difference between codominant and dominant major locus models may be
evaluated by chi-square. By contrast, the dominant and recessive models
may not be compared in this way. A convenient means of evaluating two
models that are not nested is to use the Akaike Information Criterion
[Akaike, 1974], defined as $AIC = -2L + 2K$, where L is the natural log
likelihood and K is the number of parameters estimated. The 'best' model
is the one with the minimum AIC.

Likelihood methods for genetic analysis of pedigree data have the
advantage that they can be used not only to evaluate alternative models
of disease etiology, but also to obtain estimates of gene frequencies,
penetrances and other parameters. A precaution is that ascertainment
biases and various other factors can lead to incorrect parameter estimates
and incorrect conclusions about the importance of major locus effects.
Among the factors that can lead to misinterpretation of results of segregation analysis are the presence of undetected age or sex effects, population
and etiologic heterogeneity, or shared familial environmental factors that
contribute to disease susceptibility.

Several attempts have been made to overcome these limitations of
segregation analysis. In 1983, Lalouel et al. proposed a unified model of
segregation analysis which is of value in helping to avoid false inferences
of a major locus effect. In this approach, a general model with arbitrary
transmission probabilities is compared statistically with (1) an 'environmental' model with a major effect that is not transmitted from parent
to offspring, and (2) a Mendelian transmission model. Two conditions
must be met before the presence of a major locus is inferred: the environmental model must be rejected, and the Mendelian transmission
model must not be rejected. More recently, Bonney [1986, 1987] has
proposed an alternative to the normal threshold model for genetic analysis of family data. This approach, which involves the use of regressive

logistic models, provides better methods for incorporating environmental variables that may influence disease susceptibility and for estimating the effects of these variables.

The coronary heart disease literature contains few examples of the application of segregation analysis to detecting major loci for disease per se. There are three basic reasons for this omission. First, coronary heart disease often has a late age of onset and therefore any estimate of disease liability for younger family members is likely to be unreliable. Second, the disease endpoint (however it is defined) is the result of an interacting sequence of metabolic events, only some of which may be directly related to a gene product. Finally, analysis of quantitative traits is considerably more powerful than analysis of dichotomized (affected or unaffected) data. In fact, many analyses of coronary heart disease families have focused on quantitative risk factors in both affected and unaffected family members. In any one study, individuals may be categorized as affected because (1) they have angiographically defined coronary heart disease; (2) they have survived a myocardial infarction; or (3) they have a specified dyslipoproteinemia. Evidence for major disease loci obtained by segregation analysis of quantitative risk factors is described below, in the discussion of segregation analysis of quantitative traits.

Detecting Major Loci That Affect Quantitative Risk Factors

The many quantitative risk factors for atherosclerosis (such as serum lipoprotein concentrations) are prime candidates for statistical genetic analysis. The same statistical approaches that are used to detect major disease susceptibility loci (for example, association analysis, linkage analysis, and segregation analysis) can be applied to quantitative traits such as serum concentrations of lipoproteins and apolipoproteins, traits that may be easier to define than is disease status, and that are presumed to be much closer to the level of the gene. For quantitative traits, there are many possible phenotypes, and because the phenotype of each individual is a numerical value and not simply the presence or absence of a trait, likelihood methods of analysis provide more information about the contribution of major loci. The loci that have a major effect on these risk factors may have only small effects on overall disease susceptibility, but identification of many such genes can contribute substantially to understanding of the genetic basis of disease.

An example of the value of quantitative data is provided by analyses of Lp(a), an apolipoprotein associated with apo B in LDL particles. In early analyses by several investigators, this trait was dichotomized as

Lp(a) positive or negative. Linkage analysis [Namboodiri et al., 1977] yielded a suggestive lod score of 2.32 for linkage to esterase D, now known to be located on chromosome 13. Segregation analyses revealed only marginal evidence of a major locus in one study [Morton et al., 1978a], but strong evidence in another [Iselius et al., 1981]. Subsequently, segregation analysis of quantitative levels of Lp(a) [Hasstedt et al., 1983] revealed a dominant major allele for high Lp(a) levels, with a polygenic background. It remains to be determined whether this gene is the structural locus for apolipoprotein a, which is linked to the plasminogen locus on chromosome 6 [Weitkamp et al., 1988].

A variety of statistical methods are available for detecting the contribution of major loci to quantitative traits. As in the case of dichotomous traits, these methods involve a comparison of distributions of the trait observed in families and in populations, with the distributions that are expected under various genetic hypotheses. Some methods are aimed at detecting the effects of specific loci (such as candidate genes) on disease susceptibility or quantitative risk factors, while others aim to detect the influence of as yet unidentified major genes. Among those methods that may be used to identify the contribution of specific genes or linkage groups are association analysis and linkage analysis. Methods used to detect the effects of unspecified major loci on quantitative traits include not only segregation analysis, but also a variety of other methods that utilize either family or population data. These include analysis of the population distribution of the trait (commingling analysis), and analysis of the variance in the trait within and between sibships or other groups of relatives (for example, sibship variance tests). Nonparametric methods (such as structured exploratory data analysis), which examine the pattern of similarities in trait values among family members, also have been proposed. These methods are briefly described below.

Association Analysis

Association analysis of quantitative traits involves estimation of the contribution of specific marker alleles to variation in trait levels in populations and in pedigrees. Evidence for population associations between polymorphic markers and quantitative risk factors has been sought in numerous studies. The usual approach is to examine mean values for the risk factor among individuals of different genotypes at the marker locus. For example, Law et al. [1986] found an association between a DNA polymorphism of the apo B locus and altered serum concentrations of triglyceride. Recently, methods have been developed for detecting associations between markers and quantitative traits in data from nuclear families [Boerwinkle et al., 1986] and large pedigrees [George and Elston, 1987]. The goal of these methods is not only to identify specific polymorphic loci that contribute to variation in the level of a quantitative trait, but also to estimate the magnitude of the effects of specific marker alleles. The loci of interest may be either random biochemical or DNA markers or, more typically, candidate genes.

In the approach of George and Elston [1987], data are adjusted for covariates such as age and sex. Familial correlations among individuals in the pedigree, as well as departures from normality, are taken into account. In an application of this approach, George et al. [1987] analyzed a single large pedigree with a high prevalence of heart disease. They found associations between the A antigen of the ABO blood group and both total serum cholesterol (TSC) and LDL cholesterol (LDL-C).

The 'measured genotype' approach of Boerwinkle et al. [1986] also may be used to evaluate the effects of specific alleles on variation of the trait in question. In this method, the average effect of each allele is computed as the difference in trait value between individuals who carry the allele and the population mean. The contribution of the locus to quantitative variation of the trait is the ratio of the variance among the mean trait values for the separate genotypes to the total phenotypic variance. An example of this approach [reviewed in Davignon et al., 1988] is the demonstration that allelic variation at the apolipoprotein E structural locus accounts for approximately 7% of the variance in TSC concentration. The $\epsilon 2$ allele is associated with reduced TSC, LDL-C, and LDL-apo B; and the $\epsilon 4$ allele, with increased concentrations of these lipoprotein variables [Sing and Davignon, 1985]. The effects of allelic variation at apolipoprotein loci on susceptibility to coronary heart disease has been pursued in association studies, in which the frequencies of various apolipoprotein alleles in individuals with dyslipoproteinemias or with coronary artery disease are compared with allele frequencies in controls.

Quantitative Linkage Analysis

The demonstration of cosegregation of a quantitative trait and a marker locus provides strong evidence for the existence of a major locus effect on the trait. In 1938, Penrose proposed a method for detecting linkage between a marker locus and a quantitative trait using data from sib pairs. Haseman and Elston [1972] extended this method to include marker information on parents. Their test is based upon a comparison of the estimated proportion of marker alleles identical by descent in the sib pair, and the similarity in their quantitative trait values.

Likelihood methods also can be used for quantitative linkage analysis. The approach is similar to that used for dichotomous traits. The likelihood of observing a specific pattern of marker genotypes and trait values in a set of family data is computed for a variety of genetic models. The use of LIPED for quantitative linkage analysis requires values for allele frequencies and phenotypic means and variances. With package programs such as PAP, recombination frequencies between the marker locus and the hypothesized trait locus can be estimated, as well as values for allele frequencies and other genetic parameters. The sib pair test and a test utilizing LIPED have been evaluated by Kammerer and MacCluer [1985].

Amos et al. [1987] have used both sib pair and likelihood methods to examine linkage relationships between a battery of polymorphic markers and serum concentrations of HDL-C, LDL-C, apo AI, and apo B. Although results of segregation analysis were consistent with major gene effects on apo B and HDL-C levels, there was no evidence for linkage of these major genes to any of the 31 markers. Evidence for major locus effects also was found for the ratio of HDL-C to apo AI, and the ratio of LDL-C to apo B. Possible linkage between haptoglobin and the HDL-C/apo AI ratio was indicated, with a lod score of 1.72 at $\theta = 0.05$.

Commingling Analysis

If a quantitative trait is determined by many random (nonfamilial) environmental and additive polygenic factors, each with only a small effect, then the population distribution of the trait is expected to be normal. In fact, however, many quantitative traits exhibit nonnormal distributions. A common cause of departure from normality is underlying heterogeneity.

Population heterogeneity for a quantitative trait can be produced by environmental or genetic factors. It can arise as an artifact if crucial elements of population structure (for example, ethnic heterogeneity) are ignored. Whatever its cause, the result is a population distribution consisting of a mixture of underlying distributions. If a major locus contributes to variation in a quantitative trait, then the population distribution consists of a mixture of distributions, one for each major locus genotype. Statistical methods have been developed to determine whether the observed population distribution of a quantitative trait is better approximated by a mixture (or commingling) of distributions than by a single distribution. The application of commingling analysis to lipid variables is discussed by Morton et al. [1977].

Commingling analysis is a useful preliminary method for obtaining evidence of major locus effects, and has the advantage that no family data are required. However, the demonstration that a population distribution is best fitted by a mixture of two or more distributions is not sufficient to conclude the existence of a major locus effect. Any heterogeneity within a population, whether genetically or environmentally determined, can yield evidence of commingling. Likewise, if the population distribution is skewed, then two or more distributions may provide a better fit than one. There are many possible causes of skewness, including unequal contributions of different polygenic or environmental factors, or dominance at some polygenic loci. By the use of an appropriate transformation, skewed distributions often can be normalized prior to commingling analysis, thus reducing the risk of falsely inferring two or more distributions.

Sibship Variance Tests

If polygenic and enviromental factors are the only determinants of the level of a quantitative trait, then the variance in trait values among siblings is not a function of the trait values of the parents. On the other hand, if there is a major locus with a relatively rare allele determining elevated trait values, then the within-sibship variance is a function of parental values. In particular, siblings whose parents have low trait values are expected to have a smaller within-sibship variance than do siblings from parents with high values. This relationship has been used as a basis of statistical tests for major locus effects on quantitative traits [see, for example, Fain, 1978]. Disadvantages of this approach are that it requires large numbers of sibships, and that it does not yield estimates of allele frequencies, penetrances, and other genetic parameters. However, because sibship variance methods are computationally simple and permit statistical tests of significance, they are frequently used as preliminary tests. The use of sibship variance tests in analysis of lipid levels is described by Green et al. [1984], who also discuss numerous methodological issues in the use of such methods.

Structured Exploratory Data Analysis

Many of the statistical tests described so far are based upon normal distribution theory. Because the assumption of underlying normality may be incorrect, nonparametric methods, which make no distributional assumptions, have been proposed. Among these methods is structured exploratory data analysis (SEDA), developed by Karlin et al. [1981a]. The SEDA approach uses a number of indices and measures to explore the similarity of trait values among family members. The indices include the midparent-child correlation coefficient (MPCC), the major gene index (MGI), and the offspring-between-parents function (OBP).

The MPCC compares the variation among trait values of offspring with the variation among midparent values (the average of the parental values in each family). The MPCC is expected to be near zero for a trait with no genetic determination and between approximately 0.0 and 0.7 for a trait influenced by a major locus. The MGI compares the difference between offspring and midparent values with the difference between offspring and parental values. The expectation is that these two differences will be approximately the same for a trait determined only by random environmental factors. In contrast, for traits influenced by a major locus, the difference between offspring and midparent value will be greater than the offspring-parent difference, and for polygenic traits, it will be less. The OBP gives the proportion of offspring trait values that fall within specified distances of the midparental value. Comparisons of the shape of the OBP curve are made standard reference curves for traits that are determined primarily by environmental factors, polygenic effects, or a major locus.

The power of SEDA and sibship variance tests to detect major locus effects has been investigated by MacCluer and Kammerer [1984a, b]. Karlin et al. [1981b] have applied SEDA to the analysis of lipid and lipoprotein quantities in the Stanford LRC Family Study data. The advantages of the SEDA appproach are that the indices are simple to compute, and they permit the investigator to explore the patterns of trait values in families, in ways that are sometimes difficult with computationally more demanding methods. Important disadvantages are that no statistical tests of significance are possible, and no estimates of allele frequencies or other parameters are generated. Like sibship variance tests, SEDA may be a useful preliminary step in the search for major locus contributions to quantitative traits.

Segregation Analysis

The approach used in segregation analysis of quantitative risk factors is similar to that used for dichotomous traits. The likelihoods of observing a given set of pedigree data under two different genetic models are compared using the likelihood ratio criterion (-2 times the difference between the natural log likelihoods for the two models). As for dichotomous traits, the Akaike Information Criterion (AIC) can be used for comparison of two models that are not nested. Estimates are obtained for various genetic parameters, for example, the frequencies of the major locus alleles, the mean levels of the quantitative trait for each major locus genotype, the standard deviation around each mean (often assumed to be the same for each major locus genotype), and the proportion of phenotypic variation around each mean that is due to polygenic effects.

Because the levels of many quantitative risk factors for atherosclerosis are affected by environmental and other nongenetic factors (such as age and sex), values often are corrected prior to segregation analysis. Alternatively, these factors may be included in the analysis as covariates. The distributions of lipoprotein and apolipoprotein quantities also may depart from normality. Skewness is a common finding in distributions of such quantitative traits. Because skewness can give false indications of a major locus effect, skewed data are usually transformed prior to segregation analysis using a transformation such as the natural logarithm. On the other hand, transformation also can remove evidence for a true major locus effect. Therefore, a common approach is to weigh the strength of evidence for a major locus in analyses of both untransformed and transformed data for different genetic models.

Segregation analysis of quantitative levels of HDL-C, LDL-C, TSC, triglycerides, and apolipoproteins has been used to search for evidence for the role of major loci in lipoprotein metabolism, and in susceptibility to atherosclerosis. Analyses have been done using data from normal human families, from families selected because they contain individuals with extreme lipoprotein phenotypes, and from families with a clinically defined abnormality (for example, survivors of myocardial infarction). Not unexpectedly, conclusions about major locus effects on these quantitative phenotypes sometimes differ according to the types of families studied. Methods that are appropriate when families are not randomly selected have been proposed by Rao and Wette [1987].

Segregation analysis of HDL-C concentrations has been carried out in many different populations: in families from the Lipid Research Clinics (LRC) Family Study, in a collection of normal Utah families, in families from the Jerusalem Lipid Research Center, in families ascertained through probands with coronary heart disease or other diseases, and in families ascertained through a proband with hyperalphalipoproteinemia or hypoalphalipoproteinemia [reviewed by MacCluer et al., 1988]. Evidence for a major locus effect on HDL-C levels was found in the families from Utah [Hasstedt et al., 1985, 1987] and Jerusalem [Friedlander et al., 1986], and in one analysis of LRC families [Namboodiri et al., 1985]. A later analysis of LRC data using the unified approach [Bucher et al., 1987] found evidence for two HDL-C distributions, but Mendelian transmission was rejected.

Analysis of four data sets ascertained through probands with coronary heart disease or other diseases revealed no HDL-C gene [Morton et al., 1978b; Hasstedt et al., 1984; Iselius et al., 1985; Moll et al., 1986]. However, evidence for a major locus influencing HDL-C concentration was found by Amos et al. [1986] in a large, multigenerational pedigree with a high prevalence of coronary heart disease and myocardial infarction, and by Hasstedt et al. [1986] in a set of Utah pedigrees selected through probands with early coronary heart disease, stroke, or hypertension. Two analyses of primary hypoalphalipoproteinemia pedigrees suggest that a major locus may be responsible for low HDL-C levels in those families [Third et al., 1984; Byard et al., 1984]. However, analyses of hyperalphalipoproteinemia have been less informative. Two studies [Iselius and Lalouel, 1982; Rao et al., 1983] failed to detect a major locus, and a third [Siervogel et al., 1980] suggested that a major locus may influence HDL-C levels in some families.

Fewer analyses of LDL-C levels have been done. Morton et al. [1978b], in the analysis of Japanese families cited above, found evidence for a rare major allele determining high LDL-C. Iselius et al. [1985] could detect no major locus effects on LDL-C in their analysis of 78 Swedish families.

Many statistical genetic analyses of total serum cholesterol levels have been carried out, either in normal families or in families ascertained through a proband with hypercholesterolemia. Most studies in the latter category found evidence for the segregation of a major allele producing high TSC. Elston et al. [1975], in a bivariate analysis of TSC and triglycerides, found support for a major dominant allele for hypercholesterolemia in a 195-member pedigree. Williams and Lalouel [1982], in an analysis of TSC levels in 158 Seattle pedigrees originally ascertained by Goldstein et al. [1973] through hyperlipidemic survivors of myocardial infarction, confirmed the existence of a dominant hypercholesterolemia allele. They also suggested that the sampling scheme used by Goldstein et al. may have produced spurious evidence for a major locus for combined hyperlipidemia.

Studies of TSC levels in normal families have resulted in different conclusions about the existence of major locus effects. Of those analyses indicating major TSC loci, some also found evidence for major locus effects on one or more lipoprotein components. Morton et al. [1978b] detected a major locus for TSC in their analysis of Japanese families, and implied that it was the same major locus as that producing high LDL-C levels. Simpson et al. [1981] identified a dominant single gene effect on high TSC in a study of the nuclear families of 1,000 Sydney blood donors. In analyses of data from Tecumseh, Michigan [Moll et al., 1979] and Rochester, Minnesota [Moll et al., 1983], evidence for a major locus affecting TSC levels was found in some families but not in others. No major locus effects on TSC levels were found by Iselius et al. [1985] in normal Swedish nuclear families.

There is little evidence for major locus effects on triglyceride levels, although many investigators have looked for such effects. The bivariate analysis of Elston et al. [1975] described above concluded that high tri-glyceride levels in their 195-member pedigree could be equally well ex-plained by a major locus or by environmental factors. Likewise, no sup-port for a major locus affecting triglycerides could be found by Iselius [1981] in an analysis of 36 kindreds with familial hypertriglyceridemia, by Williams and Lalouel [1982] in their reanalysis of the pedigrees of Gold-stein et al. [1973], nor by Iselius et al. [1985] in their investigation of 78 Swedish families. Results of a genetic analysis of VLDL, which is tri-glyceride rich, also were negative [Morton et al., 1978b].

Only a few studies have reported results of segregation analyses of apolipoprotein quantities. Analyses of apo AI levels have yielded con-trasting results. In an analysis of a large pedigree ascertained through cases of early myocardial infarction, Hasstedt et al. [1984] found no evi-dence for genetic transmission of apo AI level. Moll et al. [1986], on the other hand, showed evidence for a major locus affecting quantitative variation in apo AI in a sample of families at high risk for coronary artery disease. No evidence for a major HDL-C locus was found in either of these studies. Hasstedt et al. [1984] found no evidence for a major locus contributing to apo AII levels in a pedigree containing cases of early myocardial infarction, and polygenic heritability was estimated to be only 0.26. In 36 pedigrees ascertained for a variety of reasons, Hasstedt et al. [1987] found evidence for a major locus that explain 43% of the variance in apo B levels.

Conclusion

Statistical analysis of family and population data provides an impor-tant first step in detecting single genes that influence disease susceptibil-ity. Several of the methods described here are applicable both to complex diseases and to quantitative risk factors. However, in general, these methods have much greater power to detect major gene effects on quantitative traits than on dichotomous ones. As a consequence, studies of major locus effects on atherosclerosis have focused on statistical genetic analysis of lipoprotein and apolipoprotein concentrations rather than on analysis of disease status per se.

Not surprisingly, analyses of data from different populations sometimes yield discrepant results. Major disease susceptibility loci may be detectable in one population and not in another because of differences in environmental risk factors or genetic backgrounds, or because different ascertainment rules are used in different studies (for example, randomly selected families vs. families ascertained through an affected proband). Comparison of results among studies can be informative. Investigation of major locus effects can provide clues to mechanisms of disease susceptibility even if these effects are detectable only in some populations.

Most of these statistical methods were developed specifically for genetic analysis of data from human populations, in which family sizes tend to be small and environmental variability may obscure major locus effects. These techniques also offer a powerful tool for analysis of data from animal models, for which environments and family structures can be manipulated and controlled, and in which there may be useful genetic variability that is not found in human populations. The power of these statistical methods for studies of animal models is illustrated by recent work using data from pedigreed baboons on two diets, a basal diet and a diet high in cholesterol and saturated fat. Segregation analysis of data from 710 baboons in 23 sire families has revealed major locus effects on HDL-C concentration [MacCluer et al., 1988] and on apo AI concentration [Blangero et al., 1987] on both diets. Linkage analysis by the sib pair method has indicated possible linkage of the locus influencing apo AI concentration to the APRT (adenosine phosphoribosyl transferase) locus [Kammerer et al., 1987]. In humans, this locus is on the long arm of chromosome 16, together with the structural loci for CETP and LCAT. Further studies with the baboon model will be directed at the candidate genes in this linkage group.

In related work on pedigreed baboons, an association has been reported between alleles at the LDL receptor locus and quantitative variation in apo B [Hixson et al., submitted]. Future segregation and linkage analyses, using data on pedigreed animals exposed to two dietary regimes, will help to clarify the relationship between the LDL receptor and apo B quantities. Because of the close evolutionary relationship between this primate model and humans, these results will help to clarify mechanisms of lipoprotein metabolism in man. Results of statistical genetic analysis of data from this animal model also can be used to identify those individuals that are most likely to carry specific major genes, and thus most likely to be informative for molecular and metabolic studies. The identification of major loci in an appropriate animal model can lead to greater understanding of lipoprotein metabolism, even if those same loci are invariant in humans.

Univariate, single locus models such as those described here will continue to provide a useful approach to detecting major locus effects on coronary heart disease and its risk factors. However, such methods ignore many complexities. Most have assumed that the trait of interest is influenced by at most one major locus, and that there are no interactions between major locus, polygenic factors and environment. Many have assumed no correlations among the environments of family members. New methods that overcome some of these limitations are currently under development. These include multivariate methods, to detect the pleiotropic effects of single genes on two or more traits. These new approaches will increase the power of statistical analysis to detect major locus effects on disease susceptibility.

References

Akaike, H.: A new look at the statistical model identification. IEEE Trans. Automatic Control *19:* 716–723 (1974).

Amos, C. I.; Elston, R. C.; Srinivason, S. R.; Wilson, A. F.; Cresanta, J. L.; Ward, L. J.; Berenson, G. S.: Linkage and segregation analyses of apolipoproteins AI and B, and lipoprotein cholesterol levels in a large pedigree with excess coronary heart disease: The Bogalusa Heart Study. Genet. Epidemiol. *4:* 115–128 (1987).

Amos, C. I.; Wilson, A. F.; Rosenbaum, P. A.; Srinivasan, S. R.; Webber, L. S.; Elston, R. C.; Berenson, G. S.: An approach to the multivariate analysis of high-density lipoprotein cholesterol in a large kindred: The Bogalusa Heart Study. Genet. Epidemiol. *3:* 255–267 (1986).

Berg, K.; Heiberg, A.: Linkage studies on familial hyperlipoproteinemia with xanthomatosis: Normal lipoprotein markers and the C3 polymorphism. Birth Defects *12:* 266–270 (1976).

Blackwelder, W. C.; Elston, R. C.: A comparison of sib-pair linkage tests for disease susceptibility loci. Genet. Epidemiol. *2:* 85–97 (1985).

Blangero, J.; MacCluer, J. W.; Mott, G. E.: Genetic analysis of apolipoprotein A-I in two environments. Am. J. hum. Genet. *41:* A250 (1987).

Boerwinkle, E.; Chakraborty, R.; Sing, C. F.: The use of measured genotype information in the analysis of quantitative phenotypes in man. I. Models and analytical methods. Ann. hum. Genet. *50:* 181–194 (1986).

Bonney, G. E.: Regressive logistic models for familial disease and other binary traits. Biometrics *42:* 611–625 (1986).

Bonney, G. E.: Logistic regression for dependent binary observations. Biometrics *43:* 951–973 (1987).

Botstein, D.; White, R. L.; Skolnick, M.; Davis, R. W.: Construction of a genetic linkage map in man using restriction fragment length polymorphisms. Am. J. hum. Genet. *32:* 314–331 (1980).

Brown, M. S.; Goldstein, J. L.: A receptor-mediated pathway for cholesterol homeostasis. Science 232: 34–47 (1986).

Bucher, K. D.; Kaplan, E. B.; Namboodiri, K. K.; Glueck, C. J.; Laskarzewski, P.; Rifkind, B. M.: Segregation analysis of low levels of high-density lipoprotein cholesterol in the Collaborative Lipid Research Clinics Program Family Study. Am. J. hum. Genet. 40: 489–502 (1987).

Buetow, K. H.; Chakravarti, A.: Multipoint gene mapping using seriation. I. General methods. Am. J. hum. Genet. 41: 180–188 (1987a).

Buetow, K. H.; Chakravarti, A.: Multipoint gene mapping using seriation. II. Analysis of simulated and empirical data. Am. J. hum. Genet. 41: 189–201 (1987b).

Byard, P. J.; Borecki, I. B.; Glueck, C. J.; Laskarzewski, P. M.; Third, J. L. H. C.; Rao, D. C.: A genetic study of hypoalphalipoproteinemia. Genet. Epidemiol. 1: 43–51 (1984).

Cantor, R.; Rotter, J.: Marker concordance in pairs of distant relatives: A new method of linkage analysis for common diseases (submitted).

Chakravarti, A.; Elbein, S. C.; Permutt, M. A.: Evidence for increased recombination near the human insulin gene: Implication for disease association studies. Proc. natn. Acad. Sci. USA 83: 1045–1049 (1986).

Clerget-Darpoux, F.; Bonaïti-Pellié, C.: Epistasis effect: An alternative to the hypothesis of linkage disequilibrium in HLA associated diseases. Ann. hum. Genet. 44: 195–204 (1980).

Clerget-Darpoux, F.; Bonaïti-Pellié, C.; Hochez, J.: Effects of misspecifying genetic parameters in lod score analysis. Biometrics 42: 393–399 (1986).

Davignon, J.; Gregg, R. E.; Sing, C. F.: Apolipoprotein E polymorphism and atherosclerosis. Arteriosclerosis 8: 1–21 (1988).

Donald, J. A.; Humphries, S. E.; Tippett, P.; Noades, J. E.; Ball, S. P.: Linkage relationships of familial hypercholesterolemia and chromosome 19 markers. Human Gene Mapping 7: 156 (1984).

Elston, R. C.; Bailey-Wilson, J. E.; Bonney, G. E.; Keats, B. J.; Wilson, A. F.: SAGE – A package of computer programs to perform Statistical Analysis for Genetic Epidemiology. 7th Int. Congr. Hum. Genet., Berlin 1986.

Elston, R. C.; Namboodiri, K. K.; Glueck, C. J.; Fallat, R.; Tsang, R.; Leuba, V.: Study of the genetic transmission of hypercholesterolemia and hypertriglyceridemia in a 195-member kindred. Ann. hum. Genet. 39: 67–87 (1975).

Elston, R. C.; Namboodiri, K. K.; Go, R. C. P.; Siervogel, R. M.; Glueck, C. J.: Probable linkage between essential familial hypercholesterolemia and third complement component (C3). Birth Defects 12: 294–297 (1976).

Elston, R. C.; Stewart, J.: A general model for the genetic analysis of pedigree data. Hum. Hered. 21: 523–542 (1971).

Fain, P. R.: Characteristics of simple sibship variance tests for the detection of major loci and application to height, weight, and spatial performance. Ann. hum. Genet. 42: 109–120 (1978).

Falconer, D. S.: The inheritance of liability to certain diseases, estimated from the incidence among relatives. Ann. hum. Genet. 29: 51–76 (1965).

Ferns, G. A. A.; Galton, D. J.: Haplotypes of the human apoprotein AI-CIII-AIV gene cluster in coronary atherosclerosis. Hum. Genet. 73: 245–249 (1986).

Friedlander, Y.; Kark, J. D.; Stein, Y.: Complex segregation analysis of low levels of plasma high-density lipoprotein cholesterol in a sample of nuclear families in Jerusalem. Genet. Epidemiol. 3: 285–297 (1986).

George, V. T.; Elston, R. C.: Testing the association between polymorphic markers and quantitative traits in pedigrees. Genet. Epidemiol. 4: 193–201 (1987).

George, V. T.; Elston, R. C.; Amos, C. I.; Ward, L. J.; Berenson, G. S.: Association between polymorphic blood markers and risk factors for cardiovascular disease in a large pedigree. Genet. Epidemiol. *4:* 267–275 (1987).

Goldstein, J. L.; Schrott, H. G.; Hazzard, W. R.; Bierman, E. L.; Motulsky, A. G.: Hyperlipidemia in coronary heart disease. II. Genetic analysis of lipid levels in 176 families and delineation of a new inherited disorder, combined hyperlipidemia. J. clin. Invest. *52:* 1544–1568 (1973).

Green, P.; Owen, A. R. G.; Namboodiri, K.; Hewitt, D.; Williams, L. R.; Elston, R. C.: The Collaborative Lipid Research Clinics Program Family Study. Detection of major genes influencing lipid levels by examination of heterogeneity of familial variances. Genet. Epidemiol. *1:* 123–141 (1984).

Haseman, J. K.; Elston, R. C.: The investigation of linkage between a quantitative trait and a marker locus. Behav. Genet. *2:* 3–19 (1972).

Hasstedt, S. J.; Albers, J. J.; Cheung, M. C.; Jorde, L. B.; Wilson, D. E.; Edwards, C. Q.; Cannon, W. N.; Ash, K. O.; Williams, R. R.: The inheritance of high density lipoprotein cholesterol and apolipoproteins A-I and A-II. Atherosclerosis *51:* 21–29 (1984).

Hasstedt, S. J.; Ash, K. O.; Williams, R. R.: A reexamination of major locus hypothesis for high density lipoprotein cholesterol level using 2,170 persons screened in 55 Utah pedigrees. Am. J. med. Genet. *24:* 57–67 (1986).

Hasstedt, S. J.; Cartwright, P. E.: PAP – Pedigree Analysis Package. Tech. Rep. No. 13 (Department of Medical Biophysics and Computing, University of Utah 1979).

Hasstedt, S. J.; Kuida, H.; Ash, K. O.; Williams, R. R.: Effects of household sharing on high density lipoprotein and its subfractions. Genet. Epidemiol. *2:* 339–348 (1985).

Hasstedt, S. J.; Wilson, D. E.; Edwards, C. Q.; Cannon, W. N.; Carmelli, D.; Williams, R. R.: The genetics of quantitative plasma Lp(a): Analysis of a large pedigree. Am. J. med. Genet. *16:* 179–188 (1983).

Hasstedt, S. J.; Wu, L.; Williams, R. R.: Major locus inheritance of apolipoprotein B in Utah pedigrees. Genet. Epidemiol. *4:* 67–76 (1987).

Hedrick, P. W.: Gametic disequilibrium measures: Proceed with caution. Genetics *117:* 331–341 (1987).

Hegele, R. A.; Huang, L.-S.; Herbert, P. N.; Blum, C. B.; Buring, J. E.; Hennekens, C. H.; Breslow, J. L.: Apolipoprotein B-gene DNA polymorphisms associated with myocardial infarction. New Engl. J. Med. *315:* 1509–1515 (1986).

Hixson, J. E.; Cox, L. A.; Kammerer, C. M.; Mott, G. E.: Identification of an LDL receptor gene marker in baboons associated with altered levels of apolipoprotein B. Arteriosclerosis (submitted).

Hodge, S. E.; Spence, M. A.: Some epistatic two-locus models of disease. II. The confounding of linkage and association. Am. J. hum. Genet. *33:* 396–406 (1981).

Iselius, L.: Complex segregation analysis of hypertriglyceridemia. Hum. Hered. *31:* 222–226 (1981).

Iselius, L.; Carlson, L. A.; Morton, N. E.; Efendic, S.; Lindsten, J.; Luft, R.: Genetic and environmental determinants for lipoprotein concentrations in blood. Acta med. scand. *217:* 161–170 (1985).

Iselius, L.; Dahlen, G.; deFaire, U.; Lundman, T.: Complex segregation analysis of the Lp(a)/pre-beta$_1$-lipoprotein trait. Clin. Genet. *20:* 147–151 (1981).

Iselius, L.; Lalouel, J. M.: Complex segregation analysis of hyperalphalipoproteinemia. Metabolism *31:* 521–523 (1982).

Kammerer, C. M.; MacCluer, J. W.: Comparison of two preliminary methods of quantitative linkage analysis. Hum. Hered. *35:* 319–325 (1985).

Kammerer, C. M.; MacCluer, J. W.: Empirical power of four preliminary methods for ordering loci. Am. J. hum. Genet. (in press).

Kammerer, C. M.; MacCluer, J. W.; VandeBerg, J. L.; Mott, G. E.: Possible linkage between APRT and a major gene(s) for serum concentrations of lipoproteins and apolipoproteins in baboons. Am. J. hum. Genet. *41:* A257 (1987).

Karlin, S.; Williams, P. T.; Carmelli, D.: Structured exploratory data analysis (SEDA) for determining mode of inheritance of quantitative traits. I. Simulation studies on the effect of background distributions. Am. J. hum. Genet. *33:* 262–281 (1981a).

Karlin, S.; Williams, P. T.; Haskell, W. L.; Wood, P. D.: Genetic analysis of the Stanford LRC Family Study data. II. Structured exploratory data analysis of lipids and lipoproteins. Am. J. Epidemiol. *113:* 325–337 (1981b).

Khachadurian, A. K.: The inheritance of essential familial hypercholesterolemia. Am. J. Med. *73:* 402–407 (1964).

Kwiterovich, P. O., Jr.; Frederickson, D. S.; Levy, R. I.: Familial hypercholesterolemia (one form of familial type II hyperlipoproteinemia). A study of its biochemical, genetic and clinical presentation in childhood. J. clin. Invest. *53:* 1237–1249 (1974).

Lalouel, J. M.: Linkage mapping from pair-wise recombination data. Heredity *38:* 61–77 (1977).

Lalouel, J. M.; Rao, D. C.; Morton, N. E.; Elston, R. C.: A unified model for complex segregation analysis. Am. J. hum. Genet. *35:* 816–826 (1983).

Lander, E. S.; Botstein, D.: Homozygosity mapping: A way to map human recessive traits with the DNA of inbred children. Science *236:* 1567–1570 (1987).

Lathrop, G. M.; Lalouel, J. M.; Julier, C.; Ott, J.: Strategies for multilocus linkage analysis in humans. Proc. natn. Acad. Sci. USA *81:* 3443–3446 (1984).

Lathrop, G. M.; Lalouel, J. M.; White, R. L.: Construction of human linkage maps: Likelihood calculations for multilocus linkage analysis. Genet. Epidemiol. *3:* 29–52 (1986).

Law, A.; Powell, L. M.; Brunt, H.; Knott, T. J.; Altman, D. G.; Rajput, J.; Wallis, S. C.; Pease, R. J.; Priestly, L. M.; Scott, J.; Miller, G. J.; Miller, N. E.: Common DNA polymorphism within coding sequence of apolipoprotein B gene associated with altered lipid levels. Lancet *i:* 1301–1303 (1986).

Lee, J.; Burns, T. L.: Genetic dyslipoproteinemias associated with coronary atherosclerosis; in Pierpont, Moller, Genetics of cardiovascular disease, pp. 161–191 (Nijhoff, The Hague 1986).

Leren, T. P.; Borresen, A. -L.; Berg, K.; Hjermann, I.; Leren, P.: Increased frequency of the apolipoprotein E-4 isoform in male subjects with multifactorial hypercholesterolemia. Clin. Genet. *27:* 458–462 (1985).

Lusis, A. J.: Genetic factors affecting blood lipoproteins: The candidate gene approach. J. Lipid Res. *29:* 397–429 (1988).

MacCluer, J. W.; Falk, C. T.; Spielman, R. S.; Wagener, D. K.: Genetic Analysis Workshop II: Summary. Genet. Epidemiol. *1:* 147–159 (1984).

MacCluer, J. W.; Falk, C. T.; Wagener, D. K.: Genetic Analysis Workshop III: Summary. Genet. Epidemiol. *2:* 185–198 (1985).

MacCluer, J. W.; Kammerer, C. M.: Structured exploratory data analysis: An evaluation of the use of three SEDA statistics in assessing mode of inheritance; in Rao, Elston, Kuller, Feinleib, Carter, Havlik, Genetic epidemiology of coronary heart disease. Past, present, and future. Prog. clin. biol. Res., vol. 147, pp. 297–315 (Liss, New York 1984a).

MacCluer, J. W.; Kammerer, C. M.: Power of sibship variance tests to detect major genes; in Chakravarti, Human populations genetics: The Pittsburgh Symposium, pp. 125–141 (Van Nostrand, New York 1984b).

MacCluer, J. W.; Kammerer, C. M.; Blangero, J.; Dyke, B.; Mott, G. E.; VandeBerg, J. L.; McGill, H. C., Jr.: Pedigree analysis of HDL cholesterol concentration in baboons on two diets. Am. J. hum. Genet. 43: 401–413 (1988).

MacLean, C. J.; Morton, N. E.; Yee, S.: COMBIN: Combined analysis of genetic linkage, segregation and association under an oligogenic model. Tech. Rep. No. 1 (Population Genetics Laboratory, University of Hawaii 1983).

Moll, P. P.; Berry, T. D.; Weidman, W. H.; Kottke, B. A.: Childhood cholesterol levels: Indicators of heterogeneous etiologies of familial aggregation. Am. J. hum. Genet. 34: 187A (1983).

Moll, P. P.; Powsner, R.; Sing, C. F.: Analysis of genetic and environmental sources of variation in serum cholesterol in Tecumseh, Michigan. V. Variance components estimated from pedigrees. Ann. hum. Genet. 42: 343–354 (1979).

Moll, P. P.; Sing, C. F.; Williams, R. R.; Mao, S. J. T.; Kottke, B. A.: The genetic determination of plasma apolipoprotein A-I levels measured by radioimmunoassay: A study of high-risk pedigrees. Am. J. hum. Genet. 38: 361–372 (1986).

Morton, N. E.: Sequential tests for the detection of linkage. Am. J. hum. Genet. 7: 277–318 (1955).

Morton, N. E.; Gulbrandsen, C. L.; Rhoads, G. G.; Kagan, A.: The Lp lipoprotein in Japanese. Clin. Genet. 14: 207–212 (1978a).

Morton, N. E.; Gulbrandsen, C. L.; Rhoads, G. G.; Kagan, A.; Lew, R.: Major loci for lipoprotein concentrations. Am. J. hum. Genet. 30: 583–589 (1978b).

Morton, N. E.; MacLean, C. J.: Analysis of family resemblance. III. Complex segregation analysis of quantitative traits. Am. J. hum. Genet. 26: 489–503 (1974).

Morton, N. E.; MacLean, C. J.; Kagan, A.; Gulbrandsen, C. L.; Rhoads, G. G.; Yee, S.; Lew, R.: Commingling in distributions of lipids and related variables. Am. J. hum. Genet. 29: 52–59 (1977).

Morton, N. E.; Yee, S.; Lew, R.: Complex segregation analysis. Am. J. hum. Genet. 23: 602–611 (1971).

Namboodiri, K. K.; Bucher, K. D.; Kaplan, E. B.; Laskarzewski, P.; Glueck, C. J.; Rifkind, B. M.: Complex segregation analysis of high density lipoprotein cholesterol (HDL-C) levels in the Collaborative Lipid Research Clinics Family Study. Am. J. hum. Genet. 37: A202 (1985).

Namboodiri, K. K.; Elston, R. C.; Go, R. C. P.; Berg, K.; Hames, C.: Linkage relationships of Lp and Ag serum lipoproteins with 25 polymorphic markers. Hum. Genet. 37: 291–297 (1977).

Nevin, N. C.; Slack, J.: Hyperlipidaemic xanthomatosis. II. Mode of inheritance in 55 families with essential hyperlipidaemia and xanthomatosis. J. med. Genet. 5: 9–28 (1968).

Ordovas, J. M.; Schaefer, E. J.; Salem, D.; Ward, R. H.; Glueck, C. J.; Vergani, C.; Wilson, P. W. F.; Karathanasis, S. K.: Apolipoprotein A-I gene polymorphism associated with premature coronary artery disease and familial hypoalphalipoproteinemia. New Engl. J. Med. 314: 671–677 (1986).

Ott, J.: Estimation of the recombination fraction in human pedigrees: Efficient computation of the likelihood for human linkage studies. Am. J. hum. Genet. 26: 588–597 (1974).

Ott, J.: Linkage analysis and family classification under heterogeneity. Ann. hum. Genet. 47: 311–320 (1983).

Ott, J.; Falk, C. T.: Epistatic association and linkage analysis in human families. Hum. Genet. 62: 296–300 (1982).

Pagnan, A.; Zanetti, G.; Bonanome, A.; Biffanti, S.; Ehnholm, C.; Keso, L.; Lukka, M.: Apolipoprotein E polymorphism, serum lipids and occurrence of 'double pre-

betalipoproteinemia' (DPBL) in subjects from two different populations. Atherosclerosis 65: 23–28 (1987).

Penrose, L. S.: Genetic linkage in graded human characters. Ann. Eugen. 6: 133–138 (1938).

Rao, D. C.; Lalouel, J. M.; Suarez, B. K.; Schonfeld, G.; Glueck, C. J.; Siervogel, R. M.:
A genetic study of hyperalphalipoproteinemia. Am. J. med. Genet. 15: 195–204
(1983).

Rao, D. C.; Wette, R.: Nonrandom sampling in genetic epidemiology: Maximum likelihood
methods for multifactorial analysis of quantitative data ascertained through truncation.
Genet. Epidemiol. 4: 357–376 (1987).

Rees, A.; Shoulders, C. C.; Stocks, J.; Galton, D. J.; Baralle, F. E.: DNA polymorphism
adjacent to the human apolipoprotein AI gene: Relation to hypertriglyceridemia. Lancet i: 444–446 (1983).

Risch, N.: A new statistical test for linkage heterogeneity. Am. J. hum. Genet. 42: 353–364
(1988).

Siervogel, R. M.; Morrison, J. A.; Kelly, K.; Mellies, M.; Gartside, P.; Glueck, C. J.:
Familial hyper-alpha-lipoproteinemia in 26 kindreds. Clin. Genet. 17: 13–25 (1980).

Simpson, J. M.; Brennan, P. J.; McGilchrist, C. A.; Blacket, R. B.: Estimation of environmental and genetic components of quantitative traits with application to serum
cholesterol levels. Am. J. hum. Genet. 33: 293–299 (1981).

Sing, C. F.; Davignon, J.: Role of the apolipoprotein E polymorphism in determining normal plasma lipid and lipoprotein variation. Am. J. hum. Genet. 37: 268–285 (1985).

Smith, C. A. B.: The detection of linkage in human genetics. J. R. Statist. Soc. B 15: 153–
192 (1953).

Suarez, B. K.; VanEerdewegh, P.: A comparison of three affected-sib-pair scoring methods
to detect HLA-linked disease susceptibility genes. Am. J. med. Genet. 18: 135–146
(1984).

Third, J. L. H. C.; Montag, J.; Flynn, M.; Freidel, J.; Laskarzewski, P.; Glueck, C. J.:
Primary and familial hypoalphalipoproteinemia. Metabolism 33: 136–146 (1984).

Thompson, E. A.; Deeb, S.; Walker, D.; Motulsky, A. G.: The detection of linkage
disequilibrium between closely linked markers: RFLPs at the AI-CIII apolipoprotein
genes. Am. J. hum. Genet. 42: 113–124 (1988).

Weeks, D. E.; Lange, K.: The affected-pedigree-member method of linkage analysis. Am.
J. hum. Genet. 42: 315–326 (1988).

Weir, B. S.; Hill, W. G.: Nonuniform recombination within the human beta-globin gene
cluster (letter). Am. J. hum. Genet. 38: 776–778 (1986).

Weitkamp, L. R.; Guttormsen, S. A.; Schultz, J. S.: Linkage between the loci for the Lp(a)
lipoprotein (LP) and plasminogen (PLG). Hum. Genet. 79: 80–82 (1988).

Whitehead, A. S.; Solomon, E.; Chambers, S.; Bodmer, W. F.; Poney, S.; Fey, G.: Assignment of a structural gene for the third component of human complement to chromosome 19. Proc. natn. Acad. Sci. USA 79: 5021–5025 (1982).

Williams, W.; Lalouel, J. M.: Complex segregation analysis of hyperlipidemia in a Seattle
sample. Hum. Hered. 32: 24–36 (1982).

Wilson, S. R.: A major simplification in the ordering of linked loci. Genet. Epidemiol. (in
press).

Zannis, V. I.; Breslow, J. L.: Genetic mutations affecting human lipoprotein metabolism.
Adv. hum. Genet. 14: 125–215 (1985).

Jean W. MacCluer, PhD, Department of Genetics, Southwest Foundation for
Biomedical Research, San Antonio, TX 78284 (USA)

Lusis A J, Sparkes S R (eds): Genetic Factors in Atherosclerosis: Approaches and
Model Systems. Monogr Hum Genet. Basel, Karger, 1989, vol 12, pp 79–94

Chromosomal Organization of Genes Involved in Plasma Lipoprotein Metabolism: Human and Mouse 'Fat Maps'

Aldons J. Lusis[a]*, Robert S. Sparkes*[b]

[a] Departments of Medicine and Microbiology and Molecular Biology Institute, and
[b] Department of Medicine, Division of Medical Genetics, University of California,
Los Angeles, Calif., USA

Introduction

The levels and structures of blood lipoproteins are determined in large part by hereditary influences. Because of the association between interindividual variations of blood lipoproteins and atherosclerosis (Scanu, this volume), the characterization of these hereditary influences provides an opportunity to better understand the basis of genetic predisposition to atherosclerosis. Toward this goal, a number of laboratories are characterizing genes involved in lipoprotein metabolism and attempting to determine whether genetic variations of these genes are responsible in part for the interindividual variations of blood lipoproteins [Lusis, 1988]. Such genes, termed 'candidate genes', include those encoding apolipoproteins, lipoprotein receptors, enzymes involved in lipid metabolism, and lipid transfer proteins. Knowledge of the locations of these genes in humans is of importance for genetic studies in which the segregation patterns of these genes are examined with respect to the occurrence of various lipoprotein phenotypes. The mouse provides a usful model system for examining the genetic control of lipoprotein metabolism, since, as compared to human studies, genetic and biochemical analyses are greatly simplified and environmental influences can be controlled. Here, we have compiled chromosomal maps of genes which are involved in lipid metabolism in humans and in mice.

Methods for Gene Mapping

The recent rapid progress in gene mapping has been largely due to the convergence of a number of developments. These include improved statistical analyses combined with computer capabilities, long-term storage of cells in liquid nitrogen, identification of each chromosome by specific banding techniques, development of new phenotypic and DNA polymorphisms, and somatic cell hybridization techniques. Stimulus for gene mapping comes from the importance of this information at both the basic science and clinical levels. For example, gene mapping information is important in the following ways: increase knowledge of factors related to genetic recombination; identify factors which affect crossing over; give insight into how the expression of a gene is affected by its location; improve understanding of chromosome structure; determine whether genes related by function appear close together in the genome; identify genetic control and regulatory functions; give insight into relation of neoplasia and associated chromosome changes; determine the presence and role of pseudogenes; and evaluate the effects of heterochromatin on gene expression. In the clinical setting, gene map information: helps in the identification of genetic heterogeneity (genocopies); identifies persons at risk for genetic disease in the presymptomatic stage; helps in carrier detection in recessive conditions in which direct expression of the affected gene cannot be detected; and gives insight into pathogenesis of abnormal phenotype with chromosome abnormalities.

A number of methodologies have been developed for gene mapping. The methods described below reflect those which have been most useful and those which appear to have the best prospect for contributing further to gene mapping.

Genetic Linkage

Genetic linkage analysis involves the study of families in which a genetic trait is segregating. This was the earliest successful approach for mapping genes in humans and has experienced increased popularity with the development of restriction fragment length polymorphisms (RFLPs) of DNA. Linkage studies are based upon the fact that genes are assembled on chromosomes which carry the genes from parents to children during the formation of gametes. Mendel's law of independent segregation indicates that transmission of genes from parent to children should occur at random. However, when genes are linked, this law does not hold. Linkage is the tendency of genes on the same chromosome to segregate together and the closer genes are to each other on the same chromosome, the more frequently they will be transmitted together. This is based upon the crossing over that occurs in meiosis. Genes that are farther apart on some of the larger chromosomes may segregate randomly because of the increased likelihood that crossing over will occur between them. Genes that are located on the same chromosome are termed syntenic.

Because of the special characteristics of X-linked inheritance, a number of genes were readily mapped to the X chromosome. However, it has been more difficult to map genes to specific autosomes, in part because there are 22 pairs in humans.

Genetic linkage can be used to map genes to chromosomes if the linkage of the unknown or test trait is found to be linked to another gene which has been already assigned to an autosome. As noted above, genetic linkage is based upon the frequency that recombination occurs between the two genes of interest. Genes which are on different chromosomes or very far apart on a large chromosome, have an equal chance of staying together or undergoing recombination which gives by definition a maximum recombination of 50%. The closeness of two genes is indicated by centimorgans (cM). By convention, 1 cm represents 1% recombination. It has been estimated that in a 850-band human karyotype, each band on average contains 3 cM. Based on 3,000 cM in humans, each centimorgan contains 10^6 nucleotide base pairs. This assumes that the physical distance is the same as the recombination map distance, which may not be true in all instances.

Genetic variation is used to carry out genetic linkage studies. Frequent genetic variations are called polymorphisms, and the most commonly used in the past include red blood cell antigens, white blood cell antigens (HLA), isoenzymes and serum proteins. More recently, the DNA RFLPs have achieved the greatest prominence and utility for genetic linkage studies (see below).

In most instances of linkage analysis, it is not possible to simply count the number of crossovers that occur between two loci. Because of this, a number of indirect approaches have been developed to evaluate linkage. Morton [1955] developed the lod score method of determining the likelihood of linkage. The term 'lod' stands for 'log odds' which reflects the probability ratio. Linkage is considered established when a lod score of +3 (odds for linkage of 1,000 to 1) is attained for a particular recombination value. To facilitate the handling of the calculations, Ott [1974, 1977] developed a computer program called LIPED. As the gene map information increases, multilocus linkage analysis will be feasible, and new computer programs are being developed to handle this type of analysis (see MacCluer, this volume). An example of this has been recently reported [Donis-Keller et al., 1987] which describes a relatively coarse linkage map of DNA markers. With further development and refining of the map, it will be possible to use the map information to map any human gene, very much as has been done for the recombinant inbred strains of mice described below. Clearly this is becoming a very powerful tool for human gene mapping. This has been possible with the development of the restriction fragment length polymorphisms (RFLPs) of DNA. These RFLPs appear to occur very commonly and since they are stably inherited, they are highly useful for genetic linkage analysis.

A special case of close genetic linkage is linkage disequilibrium. This occurs when specific alleles at two different loci are found together more often than expected by chance. This generally is also seen with association of specific alleles with a given genetic type or genetic disease. However, it is important to note that association is different from genetic linkage. Linkage relates to the physical location of genetic loci and the alleles at these loci are useful only as markers for the loci. An evaluation of linkage requires the study of families, especially large families. On the other hand, association refers to a concurrence greater than predicted by chance between a specific allele and another trait which may or may not have an obvious genetic basis. The association studies require the evaluation of unrelated individuals.

Somatic Cell Hybridization

Somatic cell hybridization has been the most effective and successful means for gene mapping in humans. This is based upon the use of inter-specific somatic cell hybrids. For this analysis, normal human cells are generally fused with an established rodent cell line. As the fused (somatic cell) hybrid cells grow, human chromosomes are preferentially lost. By cloning a number of these hybrids, one will obtain a panel of different hybrids which contain different human chromosomes. By appropriate selection of these clones, it is possible to establish a panel which can be used to map to any human chromosome. By correlation of the expression of human genes, or the direct measurement of human genes (DNA), with a chromosome pattern for the clone, the genes can be mapped to specific human chromosomes. This method has been particularly powerful because it depends upon only finding differences between the species for a given gene. Many of the earlier studies mapped genes for human enzymes using this technique. More recently, the use of the DNA probes has received increasing attention. One of the main advantages of the DNA studies is that gene expression is not required to map the gene of interest. This approach can also be used to regionally map genes by starting with human cells that contain a chromosome rearrangement such as a translocation. A second technique which can be used to separate or segregate human chromosomes is chromosome sorting by flow cytometry. This has been particularly useful for the use of the DNA markers in which hybridization is carried out with DNA from the sorted chromosomes.

In situ Hybridization

With the use of DNA probes, it is possible to map a DNA probe to a human chromosome by direct hybridization to the chromosome. This method is proving to be most useful. At the present time, radioactive probes labeled with tritiated thymidine are hybridized to chromosomes after the DNA in the chromosomes has been denatured. Following appropriate exposure to photographic film, the sites of the labeling can be determined by examination under the microscope. By evaluation of a number of metaphases, it is generally possible to map the probe to a limited area on the chromosome; it has been estimated that this mapping is probably accurate to 10 cM (about 10 million bases). This technique does have some limitations: If the probe contains repetitive sequences, it may hybridize to many nonspecific chromosome sites. For good results, it is generally necessary to have a probe that is at least 500 bases long. Sometimes there is difficulty in obtaining clear chromosome banding patterns following the denaturation of the DNA. Finally, if pseudogenes exist for the probe, it may not be possible to establish which represents the functional gene.

Comparative Mapping

There has been evolutionary conservation of many syntenic genes among mammals [Nadeau and Taylor, 1984]. Therefore, information obtained about gene mapping in one species may be useful in the understanding of mapping in another species. Thus, if one finds that two genes are linked in a rodent, such as the mouse, it is very likely that same genes will be linked or syntenic in humans. For example, figure 1 shows the chromosomal organization of the genes for the major apolipoproteins in mice and humans. Clearly, all four apolipoprotein loci are homologous as judged by the conservation of linked markers.

Mapping Genes in Mice

The chromosomal locations of genes in mice can be examined through the use of appropriate somatic cell hybrids or by in situ hybridization as discussed above for human studies. In most cases, however, the organization of genes in mice is studied by recombination analysis. Classically, this has involved examination of the segregation patterns of genetic variations in backcross and F_2 progeny [Green, 1981]. Since a large number of genetic markers covering most of the mouse genome is available, it is usually possible to map genes by recombination with considerable precision. The use of wild derived inbred strains to aid in the identification of genetic variations, at either the protein or DNA level, is an important recent addition to gene mapping methodology [Geunet, 1986].

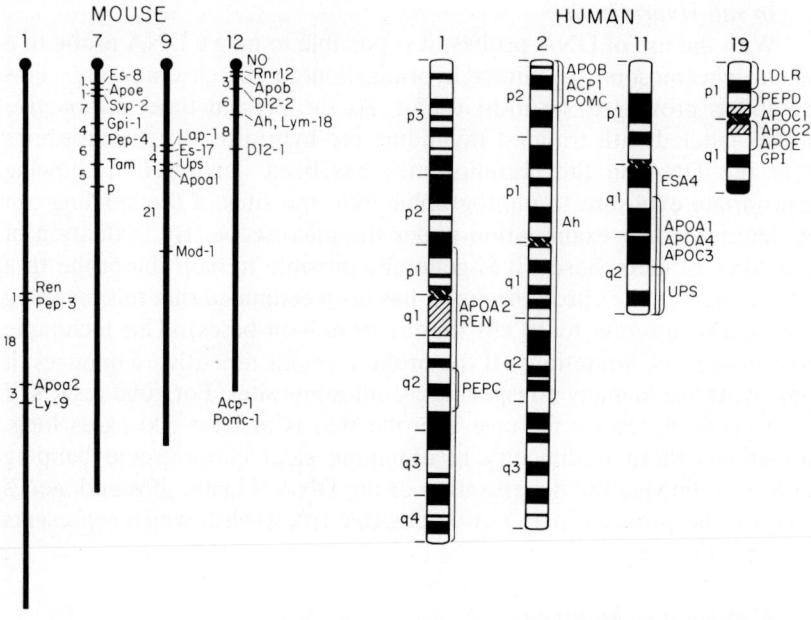

Fig. 1. Conservation of linked markers for the four loci encoding the major apolipoproteins in mice and humans. For the mouse, distances between markers are in centimorgans. For humans, regional localization of the genes is indicated by brackets. Redrawn from Lusis et al. [1987].

Traditional backcross analysis is time consuming and laborious, and much of the current gene mapping information is derived from analysis of recombinant inbred (RI) strains [Bailey, 1981; Ishida and Paigen, this volume]. RI strains are constructed by inbreeding F_2 progeny derived from two different preexisting progenitor strains. Several such RI strains are derived independently and, thus, each strain forms a stable segregant population consisting of a unique mixture of genes derived from the two parental strains. The set of RI strains permits linkage analysis, since alleles for linked genes tend to become fixed among the set of RI strains in the same combinations as in the progenitor strains, while unlinked genes become randomized with respect to one another. An important advantage of RI strains for linkage analysis is that the data for segregation of genetic markers are cumulative. Dozens of sets of RI strains have now been constructed.

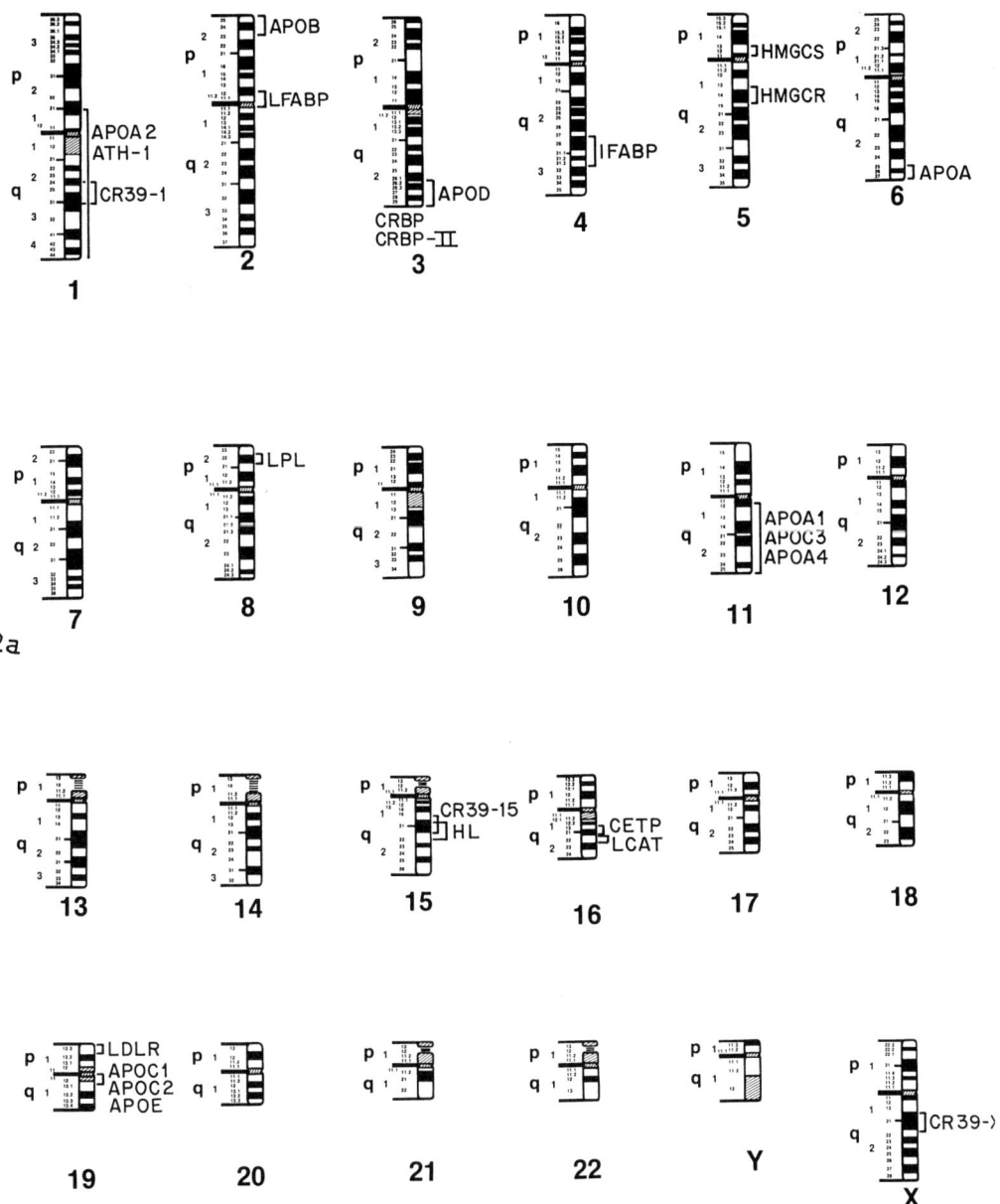

Fig. 2. The human 'fat map'. The smallest region to which the genes have been assigned is indicated by brackets. Genes which have been assigned to chromosomes but have not been regionally mapped are indicated below chromosomes.

The Human 'Fat Map'

The chromosomal locations of human genes involved in lipid metabolism are shown in figure 2. Below, we discuss briefly the characteristics of these genes, beginning with chromosome 1.

APOA2 is the structural gene for apolipoprotein AII (apo AII), one of the two major apolipoproteins of high density lipoproteins (HDL). In addition to its structural role, it may activate hepatic lipase and affect lecithin cholesterol transferase (LCAT) activity. It was mapped to the 1p21-1qter region of chromosome 1 through the use of somatic cell hybrids containing a translocation of chromosome 1 [Knott et al., 1984].

CR39-1 is one of least three loci containing genes which hybridize to cDNA for a cholesterol-responsive (CR) mRNA encoding a 39-kd protein in rat liver. As judged by size, amino acid composition and sequence homology, it appears to correspond to a prenyltransferase, a regulated enzyme of the sterol biosynthetic pathway [Clarke et al., 1987]. The location of the gene to 1q24-1q31 was determined by in situ hybridization [Lusis et al., unpubl.].

Ath-1: The *Ath-1* gene in mice determines the levels of HDL as well as susceptibility to diet-induced atherosclerosis [Paigen et al., 1987]. We have inferred the location of the presumptive corresponding human gene to human chromosome 1 on the basis of homology of flanking markers between the mouse and human genomes [Lusis et al., 1987].

APOB: This is the structural gene for apolipoprotein B, the major protein of chylomicrons, low density lipoprotein (LDL) and very low density lipoprotein (VLDL). It encodes both the high molecular weight form of the protein, apo B_{100}, incorporated by liver into VLDL, and the low molecular weight form of the protein, apo B_{48}, incorporated by intestine into chylomicrons [Powell et al., 1987]. The localization of the gene to 2p23-2p24 was by in situ hybridization as well as analysis of somatic cell hybrids [Mehrabian et al., 1986b].

LFABP is an abundant fatty acid binding protein found in liver and intestine. The proposed roles of the protein include cellular uptake of fatty acids, intracellular transport of fatty acids, and maintenance of an intracellular pool of fatty acids. The gene was localized to 2p12-2q11 by in situ hybridization and analysis of hybrid cell panels [Sweetser et al., 1987].

APOD is the structural gene for apo D, a protein constituent of HDL. The function of apo D is unclear. The gene was mapped to 3p14.2-3qter by studies with hybrid cells containing a chromosome 3 translocation. In situ hybridization yielded a major peak in the distal region of the long arm [Drayna et al., 1987].

CRBP and *CRBP-II* are genes encoding abundant cellular proteins that bind all-*trans*-retinol and vitamin A, respectively. They are members of a gene family which also includes the fatty acid binding proteins. The genes were localized to chromosome 3 by analysis of somatic cell hybrids [Demmer et al., 1987].

IFABP encodes intestinal fatty acid binding protein, a cellular protein thought to participate in the uptake, transport or metabolism of long chain fatty acids within enterocytes. It was localized to 4q28-4q31 by in situ hybridization and analysis of cell hybrids [Sweetser et al., 1987].

HMGCS is the structural gene for 3-hydroxy-3-methylglutaryl coenzyme A synthase, a highly regulated enzyme of the cholesterol biosynthetic pathway. The gene was localized to 5p14-5p12 by in situ hybridization and analysis of somatic cell hybrids with chromosome 5 deletions [Mehrabian et al., 1986a].

HMGCR encodes 3-hydroxy-3-methylglutaryl coenzyme A reductase, which is highly regulated and is probably the rate-limiting enzyme in sterol synthesis. It was localized to 5q13.3-5q14 by in situ hybridization and analysis of somatic cell hybrids with various deletions of chromosome 5 [Lindgren et al., 1985; Mohandas et al., 1986].

APOA encodes apo(a), a large glycoprotein that is disulfide linked to apo B_{100} in an LDL-like particle termed lipoprotein(a). Its normal physiological function is unknown, but high levels of the protein are correlated with increased risk of atherosclerosis. Apo(a) exhibits striking homology with plasminogen, a plasma protease, and sequence analysis suggests recent divergence of the two genes [McLean et al., 1987]. Both the apo(a) and plasminogen genes map to 6q25-6qter by in situ hybridization and analysis of cell hybrids [Frank et al., 1988].

LPL encodes lipoprotein lipase, an enzyme which functions in the hydrolysis of core triglycerides in chylomicrons and VLDL (Nilsson-Ehle et al., 1980). It is member of a gene family including hepatic lipase and pancreatic lipase (Kirchgassner et al., 1987; Wion et al., 1987). It was mapped to 8p22 by in situ hybridization and analysis of cell hybrids [Sparkes et al., 1987].

APOAI, *APOC3* and *APOA4* form a gene cluster encoding apo AI, apo CIII and apo AIV, respectively. Apo AI is the major protein of HDL and an activator of LCAT. Apo AIV is a major protein constituent of chylomicrons but its function is unclear. Apo CIII is a minor constituent of VLDL and HDL and appears to inhibit lipoprotein lipase. The organization of the gene cluster has been examined at the molecular level, and in humans it spans about 15 kb. The genes have been localized to 11q12-11qter by analysis of chromosomal translocations [Bruns et al., 1984; Cheung et al., 1984] and our preliminary in situ hybridization data suggest that the genes reside in the distal region of the long arm [Lusis et al., unpubl.].

CR39–15: This is one of three loci which hybridized with CR39 cDNA (see CR39–1). It was mapped to 15q15–15q22 by in situ hybridization [Sparkes et al., unpubl.].

HL is the structural gene for hepatic lipase, an enzyme which appears to function in the metabolism of cholesteryl ester-rich lipoproteins. Although a member of a gene family with lipoprotein lipase, the two genes are unlinked. HL was localized to 15q21 by in situ hybridization and analysis of cell hybrids [Sparkes et al., 1987].

CETP encodes cholesteryl ester transfer protein, a plasma protein which functions in the transfer of cholesteryl esters from HDL to intermediate density lipoproteins and possibly other lipoproteins. Thus, it is presumed to play an important role in 'reverse cholesterol transport', a process by which excess cholesterol is transported from peripheral tissues to liver, the only organ capable of catabolizing and secreting cholesterol. The gene was mapped to 16q12–16q21 by in situ hybridization and analysis of somatic cell hybrids [Lusis et al., 1988].

LCAT encodes lecithin-cholesterol acyltransferase, a plasma enzyme associated with HDL and involved in the esterification of HDL-cholesterol. It is important for generation of mature HDL particles and reverse cholesterol transport. The gene was localized to 16q22 by in situ hybridization and analysis of cell hybrids [Azoulay et al., 1987].

LDLR encodes the LDL receptor, which is responsible for the uptake of LDL and intermediate density lipoproteins. Defects of the LDLR are the cause of familial hypercholesterolemia and occur in about 1 in 500 individuals in the population. The gene was assigned to the short arm of chromosome 19 by analysis of a cell hybrid containing a centromeric translocation of chromosome 19 [Lusis et al., 1986]. Studies of other translocations of chromosome 19 suggest that the gene is located in the distal portion of short arm [Mohandas et al., unpubl.].

APOC1, *APOC2* and *APOE* form a tight gene complex encoding apolipoproteins CI, CII and E, respectively. Apo E serves as a ligand for both the LDL receptor and a hepatic chylomicron remnant receptor. Genetic variations of apo E have very significant effects on the levels of blood cholesterol and triglycerides in normolipidemic individuals and are involved in type III hyperlipoproteinemia, thereby contributing importantly to cardiovascular disease risk factors in the general population. Apo CI and apo CII are proteins associated primarily with triglyceride-rich lipoproteins. Apo CII is an activator of lipoprotein lipase while the function of apo CI is unclear. APOC1 and APOE have been physically linked and recombination studies indicate that APOC2 resides near this

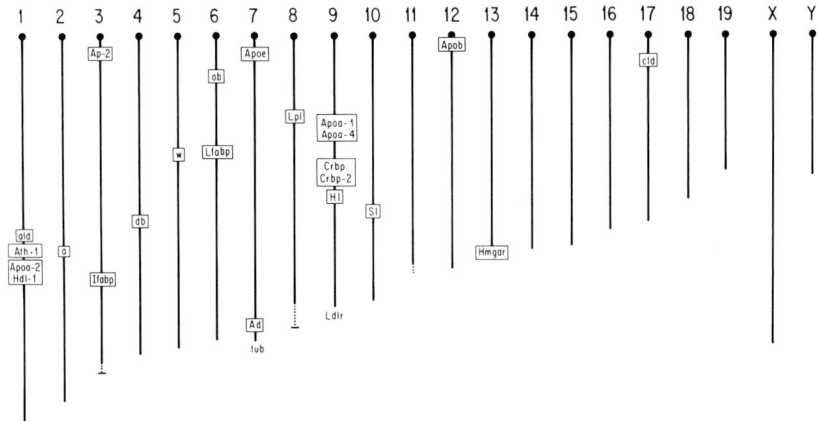

Fig. 3. The mouse 'fat map'. The approximate locations of genes, as determined by recombination with markers, are indicated. Genes which have been assigned to chromosomes but have not been regionally mapped are indicated below chromosomes.

cluster. The gene complex was assigned to 19q by analysis of cell hybrids containing a centromeric translocation of chromosome 19 and recombination studies suggest that it may reside near the centromere [Lusis et al., 1986]. Recent studies of fragmented regions of chromosome 19 indicate that the complex resides in the region 19q12–19q13.2 [Brook et al., 1987].

CR39-X is one of three loci hybridizing to CR39 cDNA (see CR39-1). It was assigned to Xq21 by in situ hybridization [Lusis et al., unpubl.].

The Mouse 'Fat Map'

Figure 3 shows the chromosomal locations of genes involved in lipid metabolism as well as various mutations which are associated with perturbations of lipid metabolism. The genes were localized primarily by linkage analysis. Below, we describe the loci, beginning with chromosome 1. [For a more extensive discussion of a number of these loci, see Ishida and Paigen, this volume].

ald is a mutation resulting in adrenocortical lipid (cholesteryl ester) depletion [Kandutsch and Coleman, 1966; Doering et al., 1973; Taylor et al., 1974].

Ath-1 is a gene which determines atherosclerosis as well as HDL levels in mice maintained on high fat diets. It maps about 6 cM proximal to the apo AII structural gene on mouse chromosome 1. The identity of the gene, and how it acts, are unknown [Paigen et al., 1987; Reue, 1985].

The gene is located in a linkage group which is conserved between mice and humans, suggesting that a corresponding gene is probably linked to APOA2 in humans [Lusis et al., 1987].

Apoa-2 is the structural gene for apo AII [Lusis et al., 1983]. Certain alleles of the gene result in altered rates of apo AII synthesis and HDL size. Another, which occurs in the senescence accelerated (SAM) mouse, results in systemic amyloidosis [Higuchi et al., 1983, 1986].

Hdl-1 affects the apolipoprotein composition and size of HDL. It probably corresponds to the apo AII structural gene [Lusis et al., 1983; Lusis, 1988].

a: The A^Y allele of the agouti locus *(a)* results in the development of obesity (due to fat cell hypertrophy) as well as yellow coat color. The allele is dominant and homozygous lethal [Herberg and Coleman, 1977].

Ap-2 encodes an adipocyte protein, referred to as aP2 or 422, that exhibits homology with members of the fatty acid binding protein gene family [Heuckeroth et al., 1987].

Ifabp encodes intestinal fatty acid binding protein [Sweetser et al., 1987].

db (diabetic) mice exhibit degranulation and eventual degeneration of β-cells, elevated plasma insulin, and marked obesity. The mutation is recessive [Coleman, 1978].

W is the dominant spotting locus. Homozygous *W* mice die early in life. Another allele of the locus is termed viable dominant spotting, W^V. Heterozygous *W/W^v* mice are anemic and lack mast cells. They exhibit variable hypertriglyceridemia, hypercholesterolemia and decreased plasma lipoprotein lipase and hepatic lipase. It has been suggested that the deficiency of the two lipases may result from improper anchoring of the enzymes to endothelial cells by heparin sulfate, since heparin is produced by mast cells [Hatanaka et al., 1986].

ob (obese) is a recessive mutation resulting in increased lipogenesis, decreased lipolysis and marked obesity [Herberg and Coleman, 1977; Coleman, 1979].

Lfabp encodes liver fatty acid binding protein [Sweetser et al., 1987].

Apoe encodes apo E [Lusis et al., 1987].

Ad (adipose) is a dominant mutation resulting in adiposity and hyperinsulinemia [Wallace and MacSwiney, 1979].

tub (tubby) mice are characterized by increased body weight and excess adipose tissue [Coleman et al., 1978].

Lpl is the structural gene for lipoprotein lipase. The gene was localized by analysis of DNA polymorphisms [LeBoeuf et al., unpubl.] and segregates with a polymorphism affecting enzyme activity in heart [Ben-Zeev et al., 1983].

Apoa-1 and *Apoa-4* are the structural genes for apolipoprotein AI and AIV, respectively. Preliminary evidence indicates that the gene for apo CIII (Apoc-3) also resides within this cluster [Lusis et al., 1983; Carrasquillo and Lusis, unpubl.].

Crbp and *Crbp-2* are genes for cellular retinal binding proteins. Genetic studies indicate that the genes are tightly linked [Demmer et al., 1987].

Hl is the structural gene for hepatic lipase [LeBoeuf et al., unpubl.]. The assignment of the gene to this chromosome 9 region is still quite tentative, since it is based on the segregation of DNA polymorphisms among very few RI strains.

Ldlr is the structural gene for the LDL receptor. It has been localized to chromosome 9 by analysis of mouse × hamster somatic cell hybrids [Heinzmann et al., unpubl.], and more recently to the proximal regions of the chromosome by analysis of DNA polymorphisms [Frank, et al., unpubl.].

Apob is the structural gene for apo B [Lusis et al., 1987].

Hmgar is the structural gene for HMG-CoA reductase. It was mapped by analysis of DNA polymorphisms [Lusis et al., unpubl.].

cld is a mutation of the *t* locus resulting in combined lipase deficiency (both lipoprotein lipase and hepatic lipase) as well as extreme hypertriglyceridemia [Paterniti et al., 1983]. It is unlinked to the genes for either lipoprotein lipase or hepatic lipase, and, therefore, presumably affects some aspect of processing common to both enzymes.

References

Azoulay, M.; Henry, I.; Tata, F.; Weil, D.; Grzeschik, K. H.; Chavez, M. E.; McIntyre, N.; Williamson, R.; Humphries, S. E.; Junien, C.: The structural gene for human lecithin: cholesterol acyl transferase (LCAT) maps to 16q22. Ann. hum. Genet. *51:* 129–136 (1987).

Bailey, D. W.: Recombinant inbred strains and bilineal congenic strains; in Foster, Small, Fox, The mouse in biomedical research, vol. I: History, genetics and wild mice, pp. 223–239 (Academic Press, New York 1981).

Ben-Zeev, O.; Lusis, A. J.; LeBoeuf, R. C.; Nikazy, J.; Schotz, M.: Evidence for independent genetic regulation of heart and adipose lipoprotein lipase activity. J. biol. Chem. *258:* 13632–13636 (1983).

Brook, J. D.; Skinner, M.; Roberts, S. H.; Rettig, W. J.; Almond, J. W.; Shaw, D. J.: Further mapping of markers around the centromere of human chromosome 19. Genomics *1:* 320–328 (1987).

Bruns, G. A. P.; Karathanasis, S. K.; Breslow, J. L.: Human apolipoprotein A-I-C-III gene complex is located on chromosome 11. Arteriosclerosis *4:* 97–102 (1984).

Cheung, P.; Kao, F. -T.; Law, M. L.; Jones, C.; Puck, T. T.; Chau, L.: Localization of the structural gene for human apolipoprotein A-I on the long arm of human chromosome 11. Proc. natn. Acad. Sci. USA *81:* 508–511 (1984).

Clarke, C. F.; Tanaka, R. D.; Svenson, K.; Wamsley, M.; Fogelman, A. M.; Edwards, P. A.: Molecular cloning and sequence of a cholesterol-repressible enzyme related to prenyltransferase in the isoprene biosynthetic pathway. Mol. cell. Biol. *4:* 3138–3146 (1987).

Coleman, D. L.: Obese and diabetes: two mutant genes causing diabetes-obesity syndromes in mice. Diabetologia *14:* 141–148 (1978).

Coleman, D. L.: Obesity genes: Beneficial effects in heterozygous mice. Science *203:* 663–665 (1979).

Coleman, D. L.; Eicher, E. M.; Southard, J. L.: Tubby. Mouse Newsl. *59:* 25 (1978).

Demmer, L. A.; Birkenmeir, E. H.; Sweetser, D. A.; Levin, M. S.; Zollman, S.; Sparkes, R. S.; Mohandas, T.; Lusis, A. J.; Gordon, J. I.: The cellular retinol binding protein II gene. Sequence analysis of the rat gene, chromosomal localization in mice and humans, and documentation of its close linkage to the cellular retinol binding protein gene. J. biol. Chem. *262:* 2458–2467 (1987).

Doering, C.; Shire, J.; Kessler, S.; Clayton, R.: Genetic and biochemical studies of the adrenal lipid depletion phenotype in mice. Biochem. Genet. *8:* 101–111 (1973).

Donis-Keller, H.; Green, P.; Helms, C., et al.: A genetic linkage map of the human genome. Cell *51:* 319 (1987).

Dragna, D. T.; McLean, J. W.; Wion, K. L.; Trent, J. M.; Drabkin, H. A.; Lawn, R. M.: Human apolipoprotein D gene: Gene sequence, chromosome localization, and homology to the α_{2u}-globulin superfamily. DNA *6:* 199–204 (1987).

Frank, S. L.; Klisak, I.; Sparkes, R. S.; Mohandas, T.; Tomlinson, J. E.; McLean, J. W.; Lawn, R. M.; Lusis, A. J.: The apolipoprotein(a) gene resides on human chromosome 6q27, in close proximity to the homologous gene for plasminogen. Hum. Genet. *79:* 352–356 (1988).

Geunet, J. L.: The contribution of wild derived mouse inbred strains to gene mapping methodology; in Potter, Nadeau, Cancro, The wild mouse in immunology: Current topics in microbiology and immunology, pp. 109–113. (Springer, Berlin 1986).

Green, M.: Gene mapping; in Foster, Small, Fox, The mouse in biomedical research, vol. I: History, genetics and wild mice, pp. 105–117. (Academic Press, New York 1981).

Hatanaka, K.; Tanichita, H.; Ishibashi-Ueda, H.; Yamamoto, A.: Hyperlipidemia in mast cell-deficient W/W^v mice. Biochim. biophys. Acta *878:* 440–445 (1986).

Herberg, L.; Coleman, D. L.: Laboratory animals exhibiting obesity and diabetes syndromes. Metabolism *26:* 59–99 (1977).

Heuckeroth, R. O.; Birkenmeier, E. H.; Levin, M. S.; Gordon, J. I.: Analysis of the tissue-specific expression, developmental regulation and linkage relationships of a rodent heart fatty acid binding protein gene. J. biol. Chem. (in press).

Higuchi, K.; Matsumura, A.; Hashimoto, R.; Honna, A.; Takeshita, S.; Hosokawa, M.; Yasohira, K.; Takeda, T.: Isolation and characterization of senile amyloid-related antigenic substance (SAS$_{SAM}$) from mouse serum. J. exp. Med. *158:* 1600–1614 (1983).

Higuchi, K.; Yonezu, T.; Kogishi K.; Matsumura, A.; Takeshita S.; Kohno, A.; Matschita, M.; Hosokawa M.; Takeda, T.: Purification and characterization of a serum amyloid-related antigenic substance (apo SAS$_{SAM}$) from mouse serum. J. biol. Chem. *25:* 12834–12840 (1986).

Kandutsch, A. A.; Coleman, D. L.: Inherited metabolic variations; in Green, Biology of the laboratory mouse, pp. 337–380 (McGraw-Hill, New York 1986).

Kirchgessner, T. G.; Svenson, K. L.; Lusis, A. J.; Schotz, M. C.: The sequence of cDNA encoding lipoprotein lipase: A member of a lipase gene family. J. biol. Chem. *262:* 8463–8466 (1987).

Knott, T. J.; Eddy, R. L.; Robertson, M. E.; Priestly, L. M.; Scott, J.; Shows, T. B.: Chromosomal localization of the human apoprotein CI gene and of a polymorphic apoprotein AII gene. Biochem. biophys. Res. Commun. *125:* 299–306 (1984).

Lindgren, V.; Luskey, R. L.; Russell, D.; Francke, U.: Human genes involved in cholesterol metabolism: Chromosomal mapping of the loci for the new density lipoprotein receptor and 3-hydroxy-3-methylglutaryl-coenzyme A reductase with cDNA probes. Proc. natn. Acad. Sci. USA *82:* 8567–8671 (1985).

Lusis, A. J.: Genetic factors affecting blood lipoprteins: The candidate gene approach. J. Lipid Res. *29:* 397–429 (1988).

Lusis, A. J.; Heinzmann, C.; Sparkes, R. S.; Scott J.; Knott, T. J.; Geller, R.; Sparkes, M. C.; Mohandas, T.: Regional mapping of human chromosome 19: Organization of genes for plasma lipid transport (APOC1, -C2, and -E and LDLR) and the genes C3, PEPD and GPI. Proc. natn. Acad. Sci. USA *83:* 3929–3933 (1986).

Lusis, A. J.; Taylor, B. A.; Quon, D. H.; Zollman, S.; LeBoeuf, R. C.: Genetic factors controlling structure and expression of apolipoproteins B and E in mice. J. biol. Chem. *262:* 7494–7604 (1987).

Lusis, A. J.; Taylor, B. A.; Wangenstein, R. W.; LeBoeuf, R. C.: Genetic control of lipid transport in mice. II. Genes controlling structure of high density lipoproteins. J. biol. Chem. *258:* 5071–5078 (1983).

Lusis, A. J.; Zollman, S.; Sparkes, R. S.; Mohandas, T.; Klisak, I.; Drayna, D.; Lawn, R. M.: Assignment of the human gene for cholesteryl ester transfer protein to chromosome 16q12–21. Genomics *1:* 232–235 (1987).

McLean, J. W.; Tomlinson, J. E.; Kuang, W. -J.; Eaton, D. L.; Chen, E. Y.; Fless, G. M.; Scanu, A. M.; Lawn, R. M.: cDNA sequence of human apolipoprotein(a) is homologous to plasminogen. Nature *330:* 132–137 (1987).

Mehrabian, M.; Calloway, K. A.; Clarke, C. F.; Tanaka, R. D.; Greenspan, M.; Lusis, A. J.; Sparkes, R. S.; Mohandas, T.; Edmond, J.; Fogelman, A. M.; Edwards, P. A.: Regulation of rat liver 3-hydroxy-3-methylglutaryl cenzyme A synthase and the chromosomal localization of the human gene. J. biol. Chem. *261:* 16249–16255 (1986).

Mehrabian, M.; Sparkes, R. S.; Mohandas, T.; Klisak, I. J.; Schumaker, V. N.; Heinzmann, C.; Zollman, S.; Ma, Y.; Lusis, A. J.: Human apolipoprotein B: Chromosomal mapping and DNA polymorphisms of hepatic and intestinal species. Somatic Cell molec. Genet. *12:* 245–254 (1986).

Mohandas, T.; Heinzmann, C.; Sparkes, R. S.; Wasmuth, J.; Edwards, P.; Lusis, A. J.: Assignment of human 3-hydroxy-3-methylglutaryl coenzyme A reductase gene to q13–q23 region of chromosome 5. Somatic Cell molec. Genet. *12:* 89–94 (1986).

Morton, N. E.: Sequential tests for the detection of linkage. Am. J. hum. Genet. *7:* 277–318 (1955).

Nadeau, J. H.; Taylor, B. A.: Lengths of chromosomal segments conserved since divergence of man and mouse. Proc. natn. Acad. Sci. USA *81:* 814–818 (1984).

Nilsson-Ehle, P.; Garfinkel, A. S.; Schotz, M. C.: Lipolytic enzymes and plasma lipoprotein metabolism. A. Rev. Biochem. *49:* 667–693 (1980).

Ott, J.: Estimation of the recombination fraction in human pedigrees: Efficient computation of the likelihood for human linkage studies. Am. J. hum. genet. *26:* 588–597 (1974).

Ott, J.: Linkage analysis with misclassification at one locus. Clin. Genet. *12:* 119–124 (1977).

Paigen, B.; Mitchell, D.; Reue, K.; Morrow, A.; Lusis, A. J.; LeBoeuf, R. C.: Ath-1, a gene determining atherosclerosis susceptibility and high density lipoprotein levels in mice. Proc. natn. Acad. Sci. USA *84:* 3763–3767 (1987).

Paterniti, J. R., Jr.; Brown, W. V.; Ginsberg, H. N.; Arzt, K.: Combined lipase deficiency (cld): A lethal mutation on chromosome 17 of the mouse. Science *221:* 167–169 (1983).

Powell, L. M.; Wallis, S. C.; Pease, R. J.; Edwards, Y. H.; Knott, T. J.; Scott, J.: A novel form of tissue-specific RNA processing produces apolipoprotein-B48 in intestine. Cell *50:* 831–840 (1987).

Reue, K. L.: The mouse as a genetic model for the study of plasma lipid transport; PhD thesis, University of California, Los Angeles (1985).

Sparkes, R. S.; Zollman, S.; Klisak, I.; Kirchgessner, T. G.; Komaromy, M. C.; Mohandas, T.; Schotz, M. C.; Lusis, A. J.: Human genes involved in lipolysis of plasma lipoproteins: Mapping of loci for lipoprotein lipase to 8p22 and hepatic lipase to 15q21. Genomics *1:* 138–144 (1987).

Sweetser, D. A.; Birkenmeier, E. H.; Klisak, I. J.; Zollman, S.; Sparkes, R. S.; Mohandas, T.; Lusis, A. J.; Gordon, J. I.: The human and rodent intestinal fatty acid binding protein genes: A comparative analysis of their structure, expression and linkage relationships. J. biol. Chem. *262:* 16060–16071 (1987).

Taylor, B.; Meier, H.; Whitten, W.: Chromosomal location and site of action of the adrenal lipid depletion gene of the mouse. Genetics *77:* S65 (1974).

Wallace, M. E.; MacSwiney, F. M.: An inherited mild middle-aged adiposity in wild mice. J. Hyg. *82:* 309–317 (1979).

Wion, K. L.; Kirchgessner,T. G.; Lusis, A. J.; Schotz, M. C.; Lawn, R. M.: Human lipoprotein lipase complementary DNA sequence. Science *235:* 1638–1641 (1987).

Aldons J. Lusis, PhD, Department of Medicine, University of California, Los Angeles, CA 90024 (USA)

Lusis A J, Sparkes S R (eds): Genetic Factors in Atherosclerosis: Approaches and Model Systems. Monogr Hum Genet. Basel, Karger, 1989, vol 12, pp 95–109

Candidate Genes for Atherosclerosis

D. J. Galton, G. A. A. Ferns [1]

Medical Professorial Unit, St. Bartholomew's Hospital, London, UK

Genes underlying the inheritance of atherosclerosis are predicted from family and twin studies. The aggregation of coronary artery disease in families has been reported by many authors since 1948 [1]. For example, Slack and Evans [2] analysed first-degree relatives of 121 men and 96 women with coronary artery disease. The increased risks of death from coronary artery disease in such relatives were 5- and 7-fold greater than in matched controls for males and females respectively. Familial clustering of coronary artery disease was noted especially for female patients. In Southern Finland, 104 out of 296 brothers of patients with coronary artery disease also had arterial disease compared to 8 out of 81 brothers of healthy controls (relative incidence 3.5 for brothers of probands with coronary artery disease [3]) A detailed analysis of the Finnish data yielded heritability estimates compatible with almost total determination of the disease by additive polygenic factors in the youngest age groups (myocardial infarction prior to the age of 46 years). In another study [4], of 207 patients who had myocardial infarcts before the age of 55 years, the highest 'risk ratios' calculated for 19 independent variables were found with a positive family history of coronary artery disease (10.5) and lesser 'risk ratios' were found with plasma cholesterol levels greater than 270 mg/dl. (4.3), cigarette smoking (4.0) and stroke in a first-degree relative (3.5). The 'risk ratio' for a family history of coronary artery disease was greater than that for individuals in the highest quintile of cholesterol levels. This observation suggests that major genetic effects are not limited to pathways of cholesterol metabolism. From this study, the heritability of coronary heart disease of early onset was calculated to be 0.63. If fami-

[1] The authors gratefully acknowledge financial support from the Wellcome Trust (to G.A.A.F.) and to the Medical Research Council (UK).

lies in which the proband had a monogenic hyperlipidaemia were eliminated, the heritability estimate remained as high as 0.56 [4]. Equally persuasive evidence comes from twin studies [5]. Concordance rates for coronary artery disease (diagnosed by angina pectoris or myocardial infarction) in monozygotic twin pairs was found to be 0.65 compared to 0.25 in dizygotic twin pairs in a Norwegian study. If twins with coronary artery disease occurring before the age of 60 years were considered alone, the concordance rates were 0.83 and 0.22 for monozygotic and dizygotic pairs respectively [6]. Of 17 twin pairs where both members had established coronary artery disease, 12 were monozygotic and 5 dizygotic, a highly significant difference of $p<0.02$. Earlier studies also found increased concordance rates for coronary artery disease in mono- compared to dizygotic twin pairs [7–9]).

There is therefore sufficient evidence from family studies to implicate genetic factors as major determinants for the development of coronary artery disease. There is less available data with regard to extracoronary disease. Most previous studies have shown correlations between the presence of coronary and extra-coronary arterial disease. For instance, in Framingham the incidence of intermittent claudication was almost 4 times higher in men and women with coronary disease than in its absence [10]. Similarly, the risk of stroke is 2–4 times higher in association with a history of angina pectoris, myocardial infarction or both than without such a history [11]. The prospective Basle Study [12] also found that men with peripheral arterial disease in the legs developed coronary artery disease about twice as often as those without. The correlations of the main manifestations of atherosclerosis in relation to risk factors strongly suggests that the same pathogenic mechanisms are involved, including genetic factors.

A useful model to consider the inheritance of atherosclerosis is the Venn diagram of figure 1 [13]. The inner circles contain subgroups of individuals within the general population (the outer circle) who possess genetic variants that confer liability or predisposition to develop atherosclerosis or hyperlipidaemia. They will not necessarily develop the disease unless they are also exposed to appropriate environmental factors, for example excess dietary intake of carbohydrate or fat (represented by the shaded sectors). Individuals exposed to the environmental factors but not possessing the genetic variants that predispose to the disease may not develop atherosclerosis.

Such a model can explain several puzzling features of the genetics of atherosclerosis. Firstly, the relatively high frequency of the disease in West European populations may be related to the fact that genes conferring susceptibility may be at no selective disadvantage under favourable

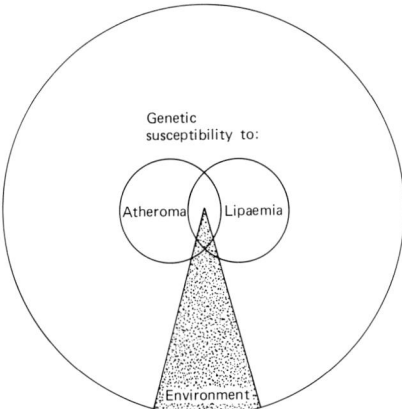

Fig. 1. A Venn diagram to illustrate the pathogenic factors for atherosclerosis. The outer circle represents a population containing subgroups (inner circles) of individuals predisposed to atheroma or hyperlipidaemia. The shaded sector represents individuals exposed to a particular environmental factor.

environmental conditions of low dietary fat intake and may be quite widespread in healthy subgroups as part of the genetic variation underlying population diversity. Some of the genetic variants predisposing to the disease must be at least as common as the disease frequency itself and may therefore constitute part of the genetic polymorphisms in populations. Secondly the variable expression of the disease in families may be due to differences in relative duration and severity of the environmental factors to which individuals are exposed or to the cumulative effects of disease-related alleles. Thirdly the 'susceptibility' loci may not code for particular proteins but may only be involved in the regulation of expression of structural genes; hence no mutant protein product would be detected in the majority of cases of premature atherosclerosis. Prior to the availability of recombinant DNA probes, there were no means of identifying genetic variants that did not result in a mutant protein. It is now possible to study the fine structure of genes suspected to be involved in the disease process, detecting variation of nucleotide sequence in introns, exons or flanking sequences; or revealing the insertion or deletion of nucleotide sequences in or adjacent to the genes of interest. Such DNA variation in flanking sequences may affect the regulation of gene expression or constitute a genetic marker for adjacent loci involved in the disease. It has been calculated that approximately 1 in every 200 base pairs may show variation in intergenic flanking sequences [14] and this is pro-

viding a wealth of new genetic markers to allow identification of adjacent
loci for analysis of their transmission in pedigrees in which atherosclerosis
segregates, or for their association in patient groups with premature
atherosclerosis [15]. Currently the latter approach has proved most informative and the rest of the chapter will deal with these aspects, i.e.
population associations of genetic variants with the occurrence of premature atherosclerosis.

Candidate Genes

Which of the 1.4 million potential genes in the human genome are
likely to be involved in the inheritance of atherosclerosis? In general,
those coding for proteins that are already known or suspected to be involved in the atherosclerosis process are worthy of study. Thus,
monogenic defects involving the LDL receptor causing familial hypercholesterolaemia give rise to premature atherosclerosis [16]. These mutations occur infrequently in Caucasian populations ($\sim 1:500$) and therefore cannot account for the common forms of atherosclerosis. However,
polymorphic DNA sequence variation within the LDL receptor gene may
lead to protein variants of the LDL receptor, modifying LDL binding
under certain environmental conditions. This may predispose to the development of hypercholesterolaemia and hence atherosclerosis. Other
proteins involved in lipid transport, particularly the apolipoproteins and
enzymes involved in lipoprotein catabolism, may also be implicated.
Genes coding for arterial wall macromolecules (such as collagen and
fibronectin), factors involved in smooth muscle cell proliferation (various
growth factors such as platelet-derived growth factor or oncogene products) and blood clotting factors such as fibrinogen [17] may also be involved (table I). At present, DNA variants at the apolipoprotein gene
clusters are being extensively studied for possible associations with premature atherosclerosis. Studies of the major gene cluster for apolipoproteins AI-CIII-AIV and the B apoprotein locus have so far produced significant results. They will now be reviewed to illustrate this general
approach.

The Apolipoprotein AI-CIII-AIV Gene Cluster

The above three genes are clustered on the long arm of chromosome
11 on a DNA segment of approximately 14 kb [18, 19]. The organization
of the cluster is presented in figure 2 and shows the following features: (1)

Table I. Candidate genes for the inheritance of atherosclerosis

Phenotype	Protein	Chromosomal location
Lipoproteins	apolipoproteins AI-CIII-AIV	11
	E-CI-CII	19
	B	2
Receptors	LDL receptor	19
	remnant receptor	
	insulin receptor	19
Enzymes	LCAT	16
	lipoprotein lipase	8
Vessel/wall proteins	fibronection	2
	collagen	17
Growth factors	platelet-derived growth factor B	22
	platelet-derived growth factor A	7
	epidermal growth factor	
	insulin	11
Coagulation factors	fibrinogen	4
	prothrombin	
	factor VII	13

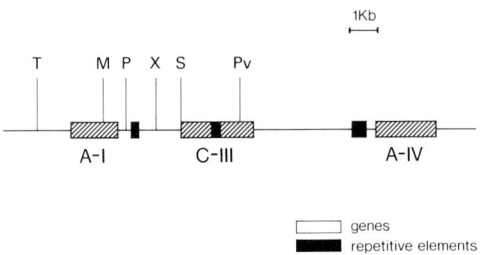

Fig. 2. The organization of the apoprotein AI-CIII-AIV gene cluster on the long arm of chromosome 11. The map shows polymorphic sites for restriction enzymes: Pu = Pvu II, S = Sst 1, X = Xmn 1, P = Pst 1, M = Msp 1 and T = Taq 1. Hatched bars are exons, solid bars are repetitive sequences.

The apo CIII gene is transcribed in the opposite direction to the apo AI and AIV genes despite being within 3 kb of the 3′ end of the apo A1 gene. (2) More than nine restriction enzyme dimorphisms along the length of this part of the genome have been described occurring within introns, exons, intergenic sequences and in flanking sequences [20, 21]. Eight studies have been performed in Caucasian populations (from the United

Kingdom, West Germany and the United States) examining the frequencies of allelic variants at these restriction sites to see if any associate with premature atherosclerosis. The results are as follow:

United Kingdom. Two groups of patients have been studied, young surviors of myocardial infarction [22] and patients with coronary or extracoronary atheroma proven by angiography [23]. In the former group, the frequency of an uncommon allele (the S2 allele) at the Sst 1 restriction site in the fourth exon of the apo CIII gene was approximately 4% in healthy controls (n = 47) compared to 21% in young survivors of myocardial infarction (n = 48). When other restriction site polymorphisms (Mspl 1 and Pst 1) were included in the analysis thereby constructing DNA haplotypes [24], it was found that one particular haplotype containing the uncommon allele at the Mspl 1 and Sst 1 sites was increased from 2% in normolipaemic controls (n = 48) to 21% in survivors of myocardial infarction (n = 47), giving a relative incidence of 12.7 (p<0.01). However, because of tight linkage disequilibrium between the alleles studied, it was not possible to identify haplotypes associated with any greater risk of premature atherosclerosis than when the Sst 1 polymorphism was considered in isolation. In a study from Edinburgh, the frequency of the S2 allele was unaltered between controls and patients with coronary heart disease but the frequency of an uncommon allele (the P2 allele) at the restriction site for Pst 1 was 6% in controls (n = 64) compared to 12% in patients with coronary heart disease (n = 79, p<0.05) [25].

Such results must be interpreted with caution, however. A common disease such as premature atherosclerosis may be particularly heterogeneous with different polygenic determinants operating in different geographical localities. The patient groups must be clearly defined and made as homogeneous as possible with regard to racial and geographical origins and clinical diagnostic features. Since atherosclerosis may have a very variable age of onset before the diagnosis can be established with certainty, the control groups may contain individuals who will go on to develop atherosclerosis at a later age. For example, the frequency of the S2 allele in healthy subjects from Scotland was reported as 18% (n = 64) compared to 4% in London (n = 47). This may represent a real difference in allelic frequencies or simply be due to differences in exclusion criteria for constituting a healthy group (i.e. presence or absence of hyperlipidaemia, diabetes or a family history of coronary artery disease, etc.). It is of interest that the S2 allele frequency in healthy Caucasian groups from Boston and Seattle are 5% (n = 66) and 6% (n = 101) respectively. This suggests that it is important to establish strict inclusion criteria for healthy control groups. Other studies [23] have used coronary arteriography to define the presence of atherosclerosis. In one report the frequency of the

S2 allele was 22% in patients with severe obstructive coronary atherosclerosis (n = 61) compared to 6% in patients with minimal disease (n =
68, p < 0.02).

Another study [26] examined the frequency of the S2 allele in extracoronary disease demonstrated by femoral or carotid angiography. In 49
Caucasian patients, 24% possessed the S2 allele compared to 4% in
healthy controls (n = 50, p < 0.01). Of 27 patients with carotid atherosclerosis, 22% had the S2 allele (p < 0.02) and in aortofemoral atherosclerosis 27% possessed the S2 allele (n = 22, p < 0.01). When normo-
lipidaemic subgroups with arterial disease were studied, the difference
in genotype frequencies persisted, 22% having the uncommon allele
(n = 32, p < 0.01), suggesting that such genetic variants have effects independent of circulating levels of plasma lipids. These and related studies
are presented in tables II and III.

United States. Studies from Boston, Seattle and New York have been
reported. In the first [27], Caucasian patients (n = 88) with severe coronary artery disease were compared to a Framingham control population
(n = 64) matched for ethnic origin and other clinical criteria were carefully standardized. The frequency of an uncommon allele revealed by the
enzyme Pst 1 at a restriction site 314 base pairs 3' to the apo A1 gene was
32% in patients compared to 4% in matched controls (p < 0.01) and 3%
in 30 subjects with no angiographic evidence for coronary artery disease,

Table II. The apo CIII mutation: frequency of the S2 allelic variants in control populations

Control groups	n	Allelic frequencies		Reference
		S1	S2	
Random MOP	37	0.96	0.04	Rees et al. [20]
Health screen clinic				
1	52	1.0	0.00	Rees et al. [45]
2	74	0.98	0.02	Ferns et al. [22]
3	56	0.98	0.02	O'Connor et al. [26]
Angio-room: normal coronary arteries	68	0.97	0.03	Rees et al. [23]
Chest clinic: random sample	35	0.99	0.01	Trembath et al. [46]
Normolipaemic controls	71	0.94	0.06	Kessling et al. [48]
Random normals	101	0.94	0.06	Deeb et al. [28]
Controls	66	0.98	0.02	Hegele et al. [29]

Results are mean allelic frequencies of a mutation involving a C-G transversion in the fourth
exon of the apolipoprotein CIII gene on the long arm of chromosome 11.

Table III. The apo CIII mutation: frequency of the S2 allelic variants in disease populations

Patient groups	n	Allelic frequencies		Reference
		S1	S2	
Hyperlipidaemic (IV/V)	28	0.80	0.20	Rees et al. [20]
Survivors of myocardial infarction (MI)	48	0.88	0.12	Ferns et al. [22]
Coronary athero-sclerosis	61	0.89	0.11	Rees et al. [23]
Peripheral atherosclerosis	49	0.86	0.12	O'Conner et al. [26]
Diabetic survivors of MI	47	0.86	0.14	Trembath et al. [46]
Hyperlipidaemia with gout	22	0.88	0.12	Ferns et al. [47]
Coronary heart disease	140	0.88	0.12	Deeb et al. [28]
Survivors of MI	66	0.96	0.04	Hegele et al. [29]

Results are mean allelic frequencies of the mutation described in the legend of table II.

giving a relative risk of coronary artery disease in individuals possessing the P2 allele of at least 10. The same rare allele was found at increased frequency in subjects with familial hypoalphalipoproteinaemia and the effect of this mutation on atherosclerosis may be mediated by lowering the levels of plasma HDL. Frequencies of alleles at other polymorphic sites at this locus were not reported. However, in a study from Seattle [28], frequencies of alleles revealed by Pst 1 and Sst 1 restriction enzymes were compared in patient groups with coronary artery disease proven by arteriography (n = 140) and random 'normals' (inclusion criteria not stated). No differences were observed in the frequency of the P2 allele; however, the frequency in the control group was 10% which is 3 times that of the Boston study. Clearly this will tend to minimize differences between patients and 'normals'. With regard to the S2 allele, frequencies were 6% (n = 101) and 12% (n = 140) in 'normals' and patients respectively (p < 0.05).

In a third study from New York [29], survivors of myocardial infarcts were examined for allelic frequencies at four polymorphic sites revealed with the enzymes Xmn 1, Msp 1, Pst 1 and Sac 1. The only significant difference was observed with the Xmn 1 polymorphism where the X2 allele frequency was lower in the patient group (n = 57) at 15% compared to 24% in controls (n = 57, p<0.05). When individuals above the age of 60 years were studied the P2 allele was significantly less frequent in patients (3 vs. 21% for controls, p<0.02). The authors concluded that DNA polymorphisms near the apo A1 gene may be significantly associated with myocardial infarction.

Federal Republic of Germany. A study from West Germany [30] examined six polymorphic sites at the apo AI-CIII-AIV gene cluster. These included sites for the enzymes Apa I, Msp I, Pst I, Ban II and Pvu II. Only one, the Pvu II site, showed an uncommon allele that was increased in frequency in patients with angiographically proven coronary atherosclerosis (n = 155) compared to healthy controls (n = 41) giving a relative incidence of 3.59 (p<0.02).

Interpretation

All seven studies thus far show restriction site polymorphisms that associate in patient groups with atherosclerosis defined either by angiography or myocardial infarctions, but the involved sites differ markedly within Caucasian populations. This may be expected if the sites are only acting as linkage markers for an atherogenic allele in the vicinity. There is also variability amongst studies regarding the frequencies of allelic variants in control populations and by affecting the comparisons with patient groups may account for some of the inconsistencies. However, some tentative conclusions may be drawn. Firstly the restriction site polymorphisms so far studied probably arise from harmless mutations and are not functioning in any way as aetiological determinants. They are possibly background (or neutral) DNA variants that differ in frequencies amongst Caucasian and other racial populations in the same way that some HLA antigen frequencies are found to vary amongst different Caucasian populations. However, the DNA polymorphisms may be acting as linkage markers for a neighbouring atherogenic mutation. There are at least two possible reasons why different neutral polymorphisms may act as linkage markers within a racial group. Firstly the putative atherogenic allele may have mutated more than once in different geographical localities and become associated with different background polymorphisms depending on which chromosome the atherogenic mutation occurred. The background polymorphisms may have since 'hitch-hiked' with the mutated atherogenic allele. An example of this is the sickle cell mutation in the β-globin gene in West Africans that occurred on a chromosome carrying an Hpa 1 polymorphism 13 kb downstream from the globin gene [31]. This Hpa 1 polymorphism acts as a linkage marker for the sickle cell mutation, but only in West African populations. Another example of this in the field of lipid metabolism is familial hypertriglyceridaemia due to mutations within the apo CII gene. In a pedigree study [32] from North Italy and Holland it was observed that different apo CII alleles revealed by a Taq 1 restriction site polymorphism tracked with affected members of each pedigree. In the North Italian family, the affected members were associated with a 3.8 kb allele; in the Dutch family by a 3.5 kb allele. The simplest explanation for these results is the apo CII mutation has occurred at least twice on different chromosomes carrying different background polymorphisms at the Taq 1 restriction site. Subsequent studies of the CII apolipoproteins have shown the occurrence of different amino-acid replacements amongst pedigrees with this form of familial hyper-triglyceridaemia [33].

The Apolipoprotein B Gene

The human apo B gene has been cloned and localized to chromosome 2 in the region of p.24 [34, 35]. It extends over 43 kb containing 29 exons and 28 introns. The distribution of introns is asymmetrical, occurring in the 5'-terminal third of the gene. The complete amino acid sequence (MW 514 kd) has been deduced from a cDNA clone [36]. A domain enriched in basic amino acids has been identified as important for the cellular uptake of cholesterol by the LDL receptor pathway. Many restriction site polymorphisms have been observed including those for the enzymes Xba 1, Eco R1 and Msp 1; and currently three studies have been published on their distribution in patients with coronary artery disease. In one study from the UK, no significant differences were observed in the allelic frequencies of the Xba 1 polymorphism at the 3' end of the apo B gene in 52 survivors of myocardial infarction and 33 healthy controls [37]. This was also observed in a second study in the USA where very similar allele frequencies were reported [28]. However, the latter group found a change in an insertional-deletional polymorphism revealed by the enzyme Msp 1 that increased in frequency in controls (n = 62) from 0.06 to 0.15 in angiographically proven patients with coronary artery disease (n = 103). This was supported in a further study [38] where the frequency of the insertional polymorphism increased from 0.142 in controls (n = 84) to 0.267 in patients (n = 84). In addition, the latter study reported an increased frequency of the rarer allele of the Xba 1 polymorphism in patients compared to controls. They concluded that the variant sites producing these polymorphisms were unlikely to be causing abnormalities of apolipoprotein B but may be simply genetic markers in linkage disequilibrium with other atherogenic mutations. None of these restriction site polymorphisms were associated with variation in lipid, lipoprotein or apolipoprotein levels and may therefore be independent risk factors for coronary artery disease. It was of interest that none of these studies revealed differences in allelic or genotype frequencies between cases and controls that were of any greater magnitude than those reported for poly-morphisms of the apo AI-CIII-AIV gene cluster. A very large difference in allele frequency between patients and controls would imply that varia-tion at that particular locus is associated with the development of coro-nary artery disease in a large proportion of cases in the general popula-tion. This may be unlikely in view of the heterogeneity of the disease and the random selection of patients in the current studies, without using any particular phenotypic feature, such as age of onset or major associated risk factors, to constitute the group.

The inconsistencies between the reports so far published may arise from variation of allelic frequencies at these polymorphic sites within genetic subgroups of a population. To interpret subsequent studies, it will be very important to ensure that cases and controls come from the same gene pool, i.e. belong to the same ethnic subgroups and preferably arise from the same geographical locality. Repetition of such studies in other racial groups and subgroups will be important to determine how wide-spread these associations and the nature of their possible phenotypic con-sequences are.

The Low Density Lipoprotein Receptor Gene and Coronary Atherosclerosis

Familial hypercholesterolaemia can be associated with various muta-tions of the LDL receptor. The resultant elevation in plasma LDL predis-poses to premature coronary atherosclerosis, and accounts for up to 6% of myocardial infarcts occurring in subjects under 60 years of age [39]. cDNA clones for the LDL receptor were first isolated in 1984 and were soon followed by the cloning and characterization of the entire genomic sequence [40]. It has an overall length of approximately 45 kb, is localized to chromosome 19 and consists of 18 exons; interestingly, 13 of these show marked sequence homologies to the genes for the C9 component of complement and epidermal growth factor. Even before the LDL receptor gene was cloned, it was realized that the molecular defect causing FH was heterogeneous [41].

Subsequent work, using recombinant DNA techniques, has con-firmed this finding and has also provided an insight into the possible mechanisms by which some of the mutations may have arisen [42]. Large deletions of the LDL receptor gene appear to be fairly common among patients with familial hypercholesterolaemia. Horsthemke et al. [43] found that approximately 6% of these patients have deletions of 1 kb or greater. The unequivocal identification of subjects with familial hypercholesterolaemia would allow early intervention and hopefully improve prognosis for macroangiography. Using a common restriction fragment length polymorphism of the LDL receptor [44], it has been possible to demonstrate the cosegregation of FH with the inheritance of the linkage marker for the gene defect in two families heralding the future prenatal diagnosis of this condition. The importance of the LDL receptor to familial hypercholesterolaemia is established but whether allelic variation at this locus contributes to polygenic hypercholesterolaemia and atherosclerosis remains to be clarified.

Future Prospects

The studies reviewed in this chapter represent the first steps in the genetic analysis using recombinant DNA probes of a complex metabolic disorder, atherosclerosis. The major new development has been the ability to detect genetic variants that do not give rise to mutant proteins in tissues yet still may be aetiological for the disease process. This has resulted from hybridization studies using recombinant DNA probes to detect genetic variation not directly affecting coding sequences. Many more loci remain to be explored other than the ones discussed above. Other candidate genes in table I will need to be analysed before a list of the genetic determinants for this disease can be tentatively adduced. It may be that different combinations of DNA polymorphisms will result in the expression of diverse types of atherosclerosis (for example, coronary, cerebral or peripheral artery disease) and family studies may provide linkage data to clarify this point. The environmental risk factors for atherosclerosis (smoking, hypertension, diabetes, etc.) combined with a knowledge of the genetic determinants will allow an accurate prediction of an individual's risk for developing the disease. This may allow early preventative measures to be instituted, preferably from early childhood and thereby prevent or delay the onset of symptoms to the last decades of life. A knowledge of the major genes involved in lipid deposition in arterial walls may also suggest new modes of therapy by possible inhibition of the function of such gene products. A rapid expansion in the understanding of this area of polygenic disease is imminent.

References

1   Yater, W. M.; Traum, A. H.; Brown, W. G.; Fitzgerald, R. P.; Geisler, M. A.; Wilcox, B. B.: Coronary artery disease in men eighteen to thirty-nine years of age. Am. Heart J. *36:* 334–372 (1948).

2   Slack, J.; Evans, K. A.: The increased risk of death from ischaemic heart disease in first-degree relatives of 121 men and 96 women with ischaemic heart disease. J. med. Genet. *3:* 239–259 (1966).

3   Rissanen, A. M.: Familial aggregation of coronary heart disease in a high incidence area (North Karelia, Finland). Br. Heart J. *42:* 294–303 (1979).

4   Nora, J. J.; Lortscher, R. H.; Spangler, R. D.; Nora, A. H.; Kimberling, W. J.: Genetic-epidemiologic study of early onset ischaemic heart disease. Circulation *61:* 503–508 (1980).

5   Berg, K.: Twin studies of coronary heart disease and its risk factors. Acta Genet. med. Gemell. *33:* 349–361 (1984).

6   Berg, K.: Genetics of coronary heart disease. Prog. med. Genet. *5:* 36–84 (1983).

7   Kahler, O. H.; Weber, R.: Zur Erbpathologie von Herz- und Kreislauferkrankungen. Z. klin. Med. *137:* 507–575 (1940).

8   Liljefors, I.: Coronary heart disease in male twins. Acta med. scand., suppl. 511 (1970).

9   De Faire, U.: Ischaemic heart disease death in discordant twins. Acta med. scand., suppl. 568 (1974).

10 Kannel, W. B.; Shurtleft, D.: The natural history of arteriosclerosis obliterans; in Brest, Gifford, Peripheral vascular disease, vol. 3, pp. 38–52 (Davis, Philadelphia 1971).

11 Kannel, W. B.; Wolf, P. A.; Verter, J.: Manifestations of coronary disease predisposing to stroke. J. Am. med. Ass. *250:* 2942–2946 (1983).

12 Widmer, L. K.; Stahelin, H. B.; Nissen, C.; Silva, A. da: Venen- und Arterienkrankheiten, und koronare Herzkrankheit bei Berufstätigen (Huber, Bern 1981).

13 Galton, D. J.: Molecular genetics of common metabolic disease (Arnold, London 1985).

14 Jeffrey, A. J.: DNA sequence variants in the γ, α, delta and beta globin genes of man. Cell *18:* 1–12 (1979).

15 Galton, D. J.: Genetic polymorphisms and atherosclerosis; in Ventura, Crepaldi, Senin, Extracoronary atherosclerosis, pp. 113–119 (Karger, Basel 1986).

16 Brown, M. S.; Kovanen, P. T.; Goldstein, J. L.: Regulation of plasma cholesterol by lipoprotein receptors. Science *212:* 628–636 (1981).

17 Meade, T. W.; Mellows, S.; Brozovic, M.: Haemostatic function and ischaemic heart disease: principal results of the Northwick Park Heart Study. Lancet *ii:* 533–537 (1986).

18 Karathanasis, S. K.; Zannis, V. I.; Breslow, J. L.: Linkage of human apolipoprotein AI and CIII genes. Nature *304:* 371–374 (1983).

19 Karathanasis, S. K.: Apolipoprotein multigene family: tandem organization of human apolipoprotein AI, CIII and AIV genes. Proc. natn. Acad. Sci. USA *82:* 6374–6378 (1985).

20 Rees, A.; Stocks, J.; Shoulders, C. C.; Galton, D. J.; Baralle, F. E.: DNA polymorphism adjacent to the human apoprotein AI gene in relation to hypertriglyceridaemia. Lancet *i:* 444–447 (1983).

21 Seilhamer, J. J.; Protter, A. A.; Frossard, P.; Levy-Wilson, B.: Isolation and DNA sequence of full length cDNA of the entire gene for human apolipprotein AI. Discovery of a new polymorphism. DNA *3:* 309–317 (1984).

22 Ferns, G. A. A.; Stocks, J.; Ritchie, C.; Galton D. J.: Genetic polymorphisms of apolipoprotein CIII and insulin in survivors of myocardial infarction. Lancet *i:* 300–304 (1985).

23 Rees, A.; Jowett, N. I.; Williams, L. G.; Stocks, J.; Vella, M. A.; Camm, J.; Galton,

D. J.: DNA polymorphisms flanking the insulin and apolipoprotein CIII genes and atherosclerosis. Atherosclerosis *58:* 269–275 (1985).

24 Ferns, G. A. A.; Galton, D. J.: Haplotypes of the human apoprotein AI-CIII-AIV gene cluster in coronary atherosclerosis. Hum. Genet. *73:* 245–249 (1986).

25 Price, W. H.; Morris, S. W.; Kitchin, A. H.: Allele frequencies at five polymorphic DNA restriction enzyme sites in the apolipoprotein AI-CIII-AIV gene cluster and coronary heart disease in a Scottish population. Clin. Sci. *75:* suppl. 16, p. 46 (1987).

26 O'Connor, G.; Stocks, J.; Lumley, J.; Galton, D. J.: A DNA polymorphism of the apolipoprotein CIII gene in extracoronary atherosclerosis. Clin. Sci. *4:* 289–292 (1988).

27 Ordovas, J. M.; Schaeffer, E. J.; Salem, D.; Ward, R. H.; Glueck, C. J.; Vergani, C.; Wilson, P. W. F.; Karathanasis, S. K.: Apolipoprotein AI gene polymorphism associated with premature coronary artery disease and familial hypoalphalipoproteinaemia. New Engl. J. Med. *314:* 671–677 (1986).

28 Deeb, S.; Failor, A.; Brown, B. G.; Brunzell, J. D.; Albers, J. J.; Motulsky, A. G.: Molecular genetics of apolipoproteins and coronary heart disease. Cold Spring Harb. Symp. quant. Biol. *51:* 403–409 (1987).

29 Hegele, R. A.; Herbert, P. N.; Blum, C. B.; Buring, J. E.; Hennekens, C. H.; Breslow, J. L.: Apolipoprotein AI and AII gene DNA polymorphisms and myocardial infarction (submitted).

30 Frossard, P. M.; Coleman, R.; Funke, H.; Assman, G.: Molecular genetics of the human apo AI-CIII-AIV gene complex; application to detection of susceptibility to atherosclerosis; in Haus, Wissler, Grünwald, Recent advances in arteriosclerosis research (Westdeutscher Verlag, Düsseldorf 1987).

31 Kan, Y. W.; Dozy, A. M.: Polymorphisms of DNA sequence adjacent to human beta-globin structural gene: relation to sickle cell mutation. Proc. natn. Acad. Sci. USA *75:* 5631–5636 (1978).

32 Humphries, S. E.; Williams, L. G.; Stahenhoef, A. F.; Baggio, G.; Crepaldi, G.; Galton, D. J.; Williamson, R.: Familial apolipoprotein CII deficiency: a preliminary analysis of the gene defect in 2 pedigrees. Hum. Genet. *65:* 151–156 (1984).

33 Baggio, G.; Gabelli, C.; Manzato, E.; Martini, S.; Previato, L.; Verlato, S.; Brewer, H. B.; Crepaldi, G.: A new apoprotein variant in 2 patients with apo CII deficiency syndrome; in Sirtori, Franceschini, Proc. NATO Advanced Research Workshop on Apolipoprotein Mutants: Apo-CII, Padova, pp. 203–210.

34 Knott, T. J.; Rall, S. C., Jr.; Innerarity, T. L.; Jacobson, S. F.; Urdea, M. S.; Levy-Wilson, B.; Powell, L. M.; Pease, R. J.; Eddy, R.; Nakai, H.; Byers, M.; Priestly, L. M.; Robertson, E.; Rall, L. B.; Betsholtz, C.; Shows, T. B.; Mahley, R. W.; Scott, J.: Human apolipoprotein B: structure of carboxy-terminal domain sites of gene expression and chromosomal localisation. Science *230:* 37–43 (1985).

35 Shoulders, C. C.; Myant, N.; Sidoli, A.; Rodriguez, J. C.; Cortese, C.; Baralle, F. E.: Molecular cloning of human LDL apolipoprotein B cDNA. Atherosclerosis *58:* 277–284 (1985).

36 Blackhart, B. D.; Ludwig, E. M.; Pierotti, R.; Carata, L.; Onasch, M. A.; Wallis, S. C.; Powell, L.; Pearse, R.; Knott, T. J.; Chu, M. L.; Mahley, R. W.; Scott, J.; McCarthy, B. J.; Levy-Wilson, B.; Structure of the human apolipoprotein B gene. J. biol. Chem. *261:* 15,364–15,367 (1986).

37 Ferns, G. A. A.; Galton, D. J.: Frequency of the Xba 1 polymorphism of the apolipoprotein B gene in myocardial infarct survivors. Lancet *ii:* 572 (1986).

38 Hegele, R. A.; Huang, L. S.; Herbert, P. N.; Blum, C. B.; Buring, J. E.; Hennekens,

C. H.; Breslow, J. L.: Apolipoprotein B gene DNA polymorhisms associated with myocardial infarction. New Engl. J. Med. *315:* 1509–1515 (1986).

39 Goldstein, J. L.; Hazzard, W. R.; Schrott, H. G.: Hyperlipidaemia in coronary heart disease. Lipid levels in 500 survivors of myocardial infarction. J. clin, Invest. *52:* 1533–1543 (1973).

40 Sudhof, T. C.; Goldstein, J. L.; Brown, M. S.; Russell, D. W.: The LDL receptor gene: a mosaic of exons shared with different proteins. Science *228:* 815–822 (1985).

41 Tolleshaug, H.; Hobgood, K. K.; Brown, M. S.; Goldstein, J. L.: The LDL receptor locus in FH – multiple mutations disrupt transport and processing of a membrane receptor. Cell *32:* 941–951 (1983).

42 Lehrman, M. A.; Schneider, W. J.; Sudhof, T. C.; Brown, M. S.; Goldstein, J. L.; Russell, D. W.: Mutations in the LDL receptor: Alu-Alu recombination deletes exons encoding transmembrane and cytoplamic domains. Science *227:* 140–146 (1985).

43 Horsthemke, B.; Dunning, A.; Humphries, S.: Identification of deletions in the human LDL receptor gene. J. med. Genet. *24:* 144–147 (1987).

44 Humphries, S. E.; Hortshemke, B.; Seed, M.; Holm, M.; Wynn, V.; Kessling, A.; Doland, J. A.; Jowett, N. I.; Galton, D. J.; Williamson, R.: A common DNA polymorphism of the LDL receptor gene and its use in diagnosis. Lancet *i:* 1003–1005 (1985).

45 Rees, A ; Stocks, J.; Sharpe, C. R.; Shoulders, C. C.; Jowett, N. I.; Baralle, F. E.; Galton, D. J.: DNA polymorphisms in the apo AI-CIII gene cluster: association with hypertriglyceridaemia. J. clin. Invest. *76:* 1090–1095 (1985).

46 Trembath, R. C.; Thomas, D. J.; Hendra, T.; Yudkin, J.; Galton, D. J.: A DNA polymorphism of the apo AI-CIII-AIV gene cluster associates with coronary heart disease in non-insulin dependent diabetes. Br. med. J. *294:* 1577–1579 (1987).

47 Ferns, G. A. A.; Lanham, J.; Galton, D. J.: Polymorphisms of the apolipoprotein AI-CIII gene cluster in subjects with hypertriglyceridaemia associated with primary gout. Hum. Genet. *75:* 121–129 (1988).

48 Kessling, A. M.; Horsthemke, B.; Humphries, S. E.: A study of DNA polymorphisms around the human apolipoprotein AI gene in hyperlipidaemic and normal individuals. Clin. Genet. *28:* 296–306 (1985).

D. J. Galton, MD, Medical Professorial Unit, St. Bartholomew's Hospital, London EC1 (UK)

Lusis A J, Sparkes S R (eds): Genetic Factors in Atherosclerosis: Approaches and Model Systems. Monogr Hum Genet. Basel, Karger, 1989, vol 12, pp 110–124

Using DNA Markers to Predict Genetic Susceptibility to Atherosclerosis

Philippe M. Frossard,[a] *Sophia Vinogradov*[b]

[a] California Biotechnology Inc., Mountain View, Calif., and
[b] Stanford University Medical Center, Stanford, Calif., USA

Introduction

The past several decades have seen a gradual but striking shift in our conceptualization of what constitutes some of the more challenging disease processes facing current clinical practice and medical research efforts. Little more than 50 years ago, medical science was preoccupied with understanding and treating the serious infectious diseases. Though often fatal, these diseases were simple to conceptualize: Koch's bacillus was the offending agent in tuberculosis; the spirochete caused syphilis. It was a paradigm in which a single etiology, the infectious agent, resulted in a single malady, the infection. Once chemicals that killed infectious agents were discovered and proven to be clinically efficacious, cure was at hand, and the effect on the practice of medicine was nothing short of revolutionary.

In recent years, however, both the diseases which represent major public health concern and our understanding of the processes which might lead to their treatment or prevention stand in stark contrast to such a simple 'one agent-one disease-one cure' paradigm. On the contrary, as the early, acute mortality of infectious diseases has for the most part been brought under control in industrialized countries,the practice of medicine and medical research has turned to a whole range of disorders which may best be described as chronic and degenerative in nature. Cardiovascular disease, adult-onset diabetes, some of the various neoplastic diseases, and dementia of the Alzheimer's type, are all examples of diseases which are insidious in onset, which often affect the middle-aged and elderly, and which cause a gradual erosion in an individual's quality of life as they consume millions of health care dollars on a national scale.

Considered one by one, these more chronic, degenerative types of diseases are often difficult to conceptualize. There is no single, clearly defined etiological agent for most of them, let alone understanding of an optimal treatment modality. For certain general categories, like cancer or cardiovascular disease, there is not even a single, clearly defined disease entity. Indeed, the three major causes of cardiovascular disease – atherosclerosis, hypertension and diabetes – are themselves heterogeneous groups of disorders, showing a plethora of clinical presentations and underlying etiological factors [Williams, 1979; Baxter et al., 1987].

In fact, these two fundamental characteristics may be the hallmark of the majority of such chronic disease processes: they are complex in their pathophysiology, and they are often multifactorial in their origins. Any of the major cardiovascular diseases is a clear case in point: both environmental factors and an underlying genetic predisposition are known to contribute to the genesis and clinical manifestations of atherosclerosis, for example. In addition, since the familial transmission of the underlying genetic predisposition in many of these disorders does not appear to be due to a single well-defined gene or gene product, such diseases are best described as both multifactorial *and* polygenic in nature.

Despite our ability to identify an individual who has been exposed to multiple environmental risk factors and who carries high genetic loading for a disease like atherosclerosis, our ability to predict the relative weight of these various factors is extremely limited. There are currently no reliable methods to calculate the probability that an individual will develop clinically significant atherosclerosis based on the relative contribution of his genetic susceptibility to environmental insults. At the same time, it is now well accepted that the manipulation of certain critical environmental factors early enough in the course of some cardiovascular disorders can often mitigate the disease process. The early identification of individuals who are, due to polygenic influences, at high risk for the development of cardiovascular disease, would thus permit a targeted preventive intervention aimed at addressing and manipulating some of the complex environmental agents that contribute to the expression of the disease.

Clinical Rationale

Cardiovascular disease is at present the leading cause of morbidity and mortality in industrialized countries. Atherosclerosis, one of the major underlying disorders contributing to cardiovascular disease, results from cholesterol deposition or atheromatous lesions in arterial walls with eventual occlusion of the vessels [see, for example, Assman, 1982].

This slow and irreversible deposition begins as early as childhood [Newman et al., 1986], although the characteristic clinical manifestations – including ischemic heart disease, cerebrovascular accidents, and peripheral vascular disease – typically do not appear before middle age. In other words, the disease process will have begun long before any clinical signs or symptoms may be observed.

Current predictive variables used to determine which individuals are at high risk for developing clinically significant atherosclerosis include a positive family history of the disease, the presence of certain 'environmental' factors (smoking history, obesity, psychosocial stressors), and serum cholesterol and triglceride levels, physiological markers that may indicate a propensity to develop atheromatous lesions. Although serum cholesterol and triglyceride levels can be measured with great accuracy, they do not appear to correlate very highly with either true susceptibility to atherosclerosis or actual presence of the diseased state. A serum cholesterol level of 310 mg/dl, for example, has been found to indicate an increased risk of only 3.1 for coronary artery disease, compared to a critical level of 180 mg/dl as recommended by the American Heart Association. Coronary angiography, or the radiologic detection of atheromatous lesions in the coronary arteries, is a more reliable diagnostic method, but it is a highly invasive procedure and cannot be employed on a routine screening basis, nor can it be applied to specific groups selected, for example, on the basis of family histories. In addition, such a diagnostic technique offers too little, too late; by the time that atheromas are radiographically detectable, the most appropriate time for vigorous preventive intervention has passed.

As discussed above, the importance of early detection of high-risk individuals is underscored by the fact that effective, albeit inconvenient, long-term preventive intervention is available to reduce some of the morbidity and mortality of atherosclerosis. Weight control, smoking cessation, lifestyle changes, and the reduction of serum cholesterol levels through strictly controlled diets or the use of cholesterol-lowering medications have all been discussed as useful preventive measures. The question remains: To whom should such rigorous prevention programs be targeted?

It has long been recognized that some forms of atherosclerosis show an inherited component, and thus individuals with a strong family history of the disorder certainly constitute a high-risk group which might benefit from rigourous preventive intervention. Because there are so many complex, interweaving variables involved in the expression of multifactorial disorders such as atherosclerosis, however, and because some forms of atherosclerosis are certainly due to polygenic effects, studies of familial inheritance by classical genetic epidemiology have for the most part failed to identify recognizable patterns of transmission for the disease. By seeking DNA markers, or restriction fragment length polymorphisms (RFLPs), associated with the presence of atherosclerosis in clinically relevant populations, investigators in the field of molecular genetics make use of the role played by various genetic components in the genesis of the disease to develop prognostic tests which identify those individuals who carry an underlying genetic predisposition to develop atherosclerotic cardiovascular disease.

Experimental Considerations

Restriction Fragment Length Polymorphisms
The procedure for studying genome organization or DNA structure involves the use of type II restriction endonucleases, naturally occurring enzymes that cleave DNA at specific permutations of nucleotides. The resulting fragments are fractionated according to size by agarose or polyacrylamide gel electrophoresis. After size separation, the fragments are transferred from the gel onto a hybridization filter [Southern, 1975], and are then hybridized to a specific cloned or synthesized labeled DNA sequence, the hybridization probe.

Nucleotide substitutions which either create or destroy a recognition site alter the distribution of fragment sizes. Nucleotide insertions, nucleotide deletions, and rearrangements of DNA sequences, also alter the distribution of fragment sizes after digestion by a restriction endonuclease. These variations in the distribution of fragment patterns among individuals constitute RFLPs. Since the nucleotide diversity in the human genome has been determined to be 0.05, and since the human genetic material contains approximately 5×10^9 base pairs (bp), about 10^7 mutations are expected to occur naturally, most of them in noncoding regions of DNA [Jeffries, 1979; Chakravarti et al., 1986]. One aim of the field of medical molecular genetics is to discover which ones among these polymorphic patterns give information about a diseased phenotype under study. DNA variations in the human genome can be arbitrarily divided into two classes. The first of these comprises DNA variations which have a direct cause-to-effect relationship with the diseased phenotype. This occurs through a variety of mechanisms, including nucleotide mutations which result in variant proteins, which modify intron-exon splicing sites and polyadenylation sites, which introduce stop codons and frame shifts, or which affect the regulation of gene expression if they are in a gene promoter region. The second class is represented by DNA variations which have no cause-to effect relationship with the diseased phenotype under study. DNA variations which belong to this class occur either out-side of coding regions, or are silent mutations; it is conceivable, however, that a silent mutation could also affect, for example, RNA stability. This stresses the difficulty to establish a cause-to-effect relationship between a mutation and a particular phenotype. The difficulty is even greater if the phenotype under study is the result of polygenic interactions.

Despite our limited knowledge of the molecular mechanisms which regulate gene expression, we can be certain that the vast majority of the nucleotide variations visualized as RFLPs belong to this latter class. They simply result from events independent of those directly responsible for the diseased phenotype under study, and the RFLP is used as a codomi-nant marker that defines a segment of chromosome on which a gene, whose product is involved in the delineation of the studied phenotype, functions in an abnormal fashion. At the stage of research where a marker-disease association is first observed, the cause of the disease is unknown. Then, by walking, jumping, or hopping along the chromosom-al segment [Weissman, 1987], one expects eventually to pinpoint which gene is involved, and how the gene defect causes the disease.

RFLPs are inherited as simple Mendelian codominant markers, and those RFLPs associated with a particular disease are occasionally referred to as gene markers for the disease.

Using Gene Markers

Gene markers can be used in two different research strategies: link-age analysis and studies of association.

(1) *Linkage analysis*: The determination of the amount of linkage between an RFLP and a given gene within a pedigree is referred to as *linkage analysis*. Linkage indicates the presence of a gene physically located in the vicinity of a marker under study; the gene may or may not segregate with the marker in family studies. The recombination rate will influence the accuracy of the information provided by linkage analysis; the use of multiple DNA probes on both sides of the gene under study can reduce the amount of type II error[1] resulting from recombination events [Martin, 1987].

Such studies are appropriate when the disease under investigation is known or felt to be monogenic in nature [Suarez and Cox, 1985]. Some

---

[1] Whenever proportions from two independent sample populations are compared, two kinds of errors are encountered and lead to false result interpretations: type I error, where a proportion difference is observed when there is in fact no difference; and type II error, when the two proportions are not observed to be different while they in fact are [see, for example, Fleiss, 1981].

forms of familial hypercholesterolemia (FH), known to be due to an autosomal-dominant defect at the LDL receptor locus, for example, are amenable to linkage analysis using RFLPs at the LDL receptor gene locus. In some families, the identification of a *Pvu*II RFLP linked to the FH allele has allowed detection of those family members who have inherited this severe form of atherosclerotic illness [Humphries et al., 1985; Leppert et al., 1986].

(2) *Association studies* delineate the amount of association between a gene marker and a clinical phenotype in a given population; they are based on a concept of statistical rather than physical proximity. Association studies, which describe the presence of a marker at a higher or lower frequency in patients when compared to disease-free controls derived from the same population, are more appropriately applied to multifactorial diseases in which the mode of inheritance is not known. They are also the only possible line of investigation for polygenic disorders, as various DNA markers at multiple gene loci can each be investigated as potential source of statistical information. It is preferable – though not always possible, particularly in the case of complex polygenic diseases – to perform confirmatory studies in large well-defined pedigrees, so that the cosegregation of these gene markers with the diseased phenotype can be established in a definitive manner.

The frequency of association between a gene marker and a 'diseased' gene will depend on variations in the frequency of the marker among different populations. Population groups of differing ethnicity, race, or geographical origin may not show the same RFLPs in association with a specific gene, particularly when mutations resulting in a 'diseased' gene arose spontaneously in various populations because of certain selective pressures. The *Hpa*I dimorphic site 3' to the human beta-globin gene associated with sickle cell anemia [Kan and Dozy, 1978] constitutes such an example.

Research Strategies

The use of RFLPs as gene markers for a disease is based on one of two research strategies:

(1) A *targeted approach* involves probing areas of the human genome that are known or felt to be involved in disorders that carry a genetic component. Since the deposition in the inner wall of arteries of serum circulating lipids is central to the process of atherosclerosis, a targeted approach for the study of molecular mechanisms involved in the development of the disease consists in studying those genes whose products are involved in lipid metabolism, transport, and clearance. In that respect,

apolipoprotein genes constitute candidate genes of choice for the study of atherosclerosis.

Indeed, apolipoproteins, complexed with lipids, form lipoproteins, whose role is to transport dietary and endogenously synthesized lipids from one organ to another [Breslow, 1985; Zannis and Breslow, 1985; Brown and Goldstein, 1986]. Besides being structural components of circulating lipoproteins, apolipoproteins are also cofactors of enzymes involved in lipid metabolism and ligands for cell surface receptors involved in lipoprotein uptake [Breslow, 1985]. It is known that dyslipoproteinemias – elevation of the atherogenic low density lipoprotein cholesterol (LDL-C) and decreased levels of the protective high density lipoprotein cholesterol (HDL-C) – increase an individual's risk of developing premature CHD. In fact, probably up to 70% of individuals suffering from premature CHD have some form of familial dyslipoproteinemia [Glueck et al., 1985].

Serum lipoprotein levels have been shown to be genetically determined. Path analysis studies, for example, have established that genetic factors account for most of the variance of HDL-C levels ($h^2 = 0.47 \pm 0.10$) and that environmental factors exert a lesser effect ($c^2 = 0.12 \pm 0.03$) [Glueck et al., 1985]. A targeted approach in the study of atherosclerosis would thus involve, first, searching for RFLPs at loci whose gene products are known to be involved in the onset of atherosclerosis, such as an apolipoprotein gene locus; and second, determining whether the frequency of some of these RFLPs is significantly different in a group of individuals with the disease and in a comparison group of subjects who are not affected.

(2) A *random, or comprehensive, approach* is based on the isolation of random DNA fragments from cDNA or genomic libraries and the use of these fragments of unknown function as hybridization probes to detect RFLPs. The degree of association between these randomly generated RFLPs and phenotypic manifestations of a disease under study are determined in a pedigree or in a sample of diseased individuals, as discussed above.

In the study of a multifactorial, polygenic disorder such as atherosclerosis, the distinction between the two approaches is not always obvious since a targeted approach can secondarily lead to a random approach. Indeed, one may choose as a targeted approach to study RFLPs at the human apo AI-CIII-AIV gene complex as they relate to, for example, hypoalphalipoproteinemia (decreased serum HDL-C levels). If an RFLP correlates with hypoalphalipoproteinemia, as does the absence of a *Pst*I dimorphic site located 0.2 kb 3' to the apolipoprotein AI (apo AI) gene [Ordovas et al., 1986][2], a reasonable explanation is that it is associated with an apo AI gene abnormality. This is a legitimate conclusion if we assume that the association is real and not due to investigation of a biased sample population. Alternatively, the observed association might be due to altered apo CIII or apo AIV gene expression. It may even be due to other genes which are in the vicinity of the apo AI-CIII-AIV gene complex locus but which have not yet been characterized. Given our limited knowledge of the mechanisms which play a role in atherosclerosis, it is conceivable that such a gene (or genes) involved in delineating the HDL phenotype could exist and not yet have been identified. A targeted approach may therefore turn into a more comprehensive approach, whereby a candidate gene used as a specific probe to detect RFLPs which are associated with a phenotype under study will indeed detect such markers, but these markers in fact indicate an as yet unsuspected role for neighboring genes in the disease.

Overview of the Work Done to Date

Molecular genetics is a new and promising approach for deciphering the etiological puzzle of atherosclerosis. Studies aimed at unraveling correlations between DNA markers and lipid disorders, myocardial infarction (MI), and coronary artery disease (CAD), have all been undertaken to date. These studies benefit from the large body of data generated by clinical investigations, data in which numerous proteins have been identified and found to be involved in the onset of atherosclerosis. Genes coding for several of these proteins have already been the focus of targeted searches for RFLPs: included are human apolipoprotein (apo) AI, AII, AIV, B, CI, CII, CIII, and E genes; the LDL receptor gene; and the insulin gene. Most investigations have focused on association studies between RFLPs and atherosclerosis, although cosegregation of RFLPs and specific defects have also been studied in several pedigrees. We will briefly review here relevant findings reported in the literature. A more extensive review on molecular genetics of apolipoproteins as they relate to atherosclerosis susceptibility has recently been published [Hegele and Breslow, 1987].

The Apo AI-CIII-AIV Gene Complex
The human apo AI-CIII-AIV gene complex spans an 18-kb region on the long arm of chromosome 11 [Karathanasis, 1985]. To date, 18 poly-

---

[2] This study is only taken as an example for the sake of the argument.

morphic sites have been reported and mapped at the apo AI-CIII-AIV gene locus [Frossard et al., submitted]; several of these sites have been studied for associations with lipid disorders and atherosclerosis.

(1) Deletion of a 300 bp Alu-like sequence 4.0 kb 5' to the apo AI gene has been associated with hypoalphalipoproteinemia [Lim et al., 1986] and increased incidence of coronary atherosclerosis in Germans [Frossard et al., 1986, 1987].

(2) Presence of an *XmnI* site 2.5 kb 5' to the apo AI gene is associated with type IIb, III, and V hyperlipoproteinemia, but not with type IIa and IV [Kessling et al., 1985]; in other reports, it is a protective marker for coronary atherosclerosis [Frossard et al., 1986, 1987] and MI [Hegele and Breslow, 1987].

(3) Absence of an *MspI* site within the third intron of the apo AI gene has been successively reported to be a high risk marker for MI [Ferns and Galton, 1986a], a protective marker for CAD [Frossard et al., 1986, 1987], and a neutral marker for MI [Hegele and Breslow, 1987].

(4) Absence of a *PstI* site 0.5 kb 3' to the apo AI gene has been found to be neutral with respect to both dyslipoproteinemia among English [Kessling et al., 1985] and MI among Bostonians [Hegele and Breslow, 1987], but associated with hypoalphalipoproteinemia and MI in another study [Ordovas et al., 1986], and with CAD among Germans [Frossard et al., 1986, 1987].

(5) Presence of an *SstI* (*SacI*) site in the 3' untranslated region of apo CIII has also led to controversial reports. It is associated with hypertriglyceridemia [Rees et al., 1983; 1985a, b; Anderson and Wallace, 1985], MI, and CAD in some studies [Ferns et al., 1985; Rees et al., 1985b; Ferns and Galton, 1986a], but not in others [Morris and Price, 1985; Frossard et al., 1986, 1987; Hegele and Breslow, 1987].

(6) Presence of a *PvuII* site in the first intron of apo CIII is a high-risk marker for CAD among Germans [Frossard et al., 1986, 1987]. DNA rearrangement in the apo AI-CIII gene region is associated with premature atherosclerosis in isolated families [Karathanasis et al., 1983, 1987].

(7) Finally, Ward et al. [1987] have reported results of pedigree analysis studies showing that the effect of the *XmnI*, *MspI*, and *PstI* RELPs of the apo AI gene locus account for 2–3% of the total population variation of serum cholesterol and triglycerides, and that these three sites are not informative markers for CAD.

The Apo B Gene

Several RFLPs have been reported at the human apo B gene locus, which is located on chromosome 2p [see Hegele and Breslow, 1987]. Of those, an *XbaI* site located in the last exon of the apo B gene has been associated successively with hypertriglyceridemia [Law et al., 1986], hypercholesterolemia [Talmud and Humphries, 1986], and MI [Hegele et al., 1986], while another report has indicated no association with CAD [Ferns and Galton, 1986b].

The Apo AII Gene

Absence of an *Msp*I dimorphic site located 0.2 kb 3' to the apo AII gene (chromosome 1q; see Hegele and Breslow [1987]) has been correlated with increased serum levels of apo AII [Scott et al., 1985] and triglycerides [Ferns et al., 1986], although two other studies failed to show evidence for association with either altered apo AII and triglyceride levels, or with MI and CAD [Schulte et al., 1986; Hegele and Breslow, 1987].

The Apo E-CI-CII Gene Complex

The apo CI, apo CII, and apo E genes lie as a complex on the long arm of chromosome 19 [see Hegele and Breslow, 1987]. Several studies on association between lipid disorders and atherosclerosis, and RFLPs detected with apo CII and apo E probes, have been reported. Three dimorphic sites detected with *Ban*I, *Nco*I, and *Taq*I and spanning a 17-kb region around the apo CII gene failed to show association with either dyslipemias or CAD and MI [Humphries et al., 1983; Assmann et al., 1986]. Absence of an *Hpa*I dimorphic site at the apo E gene locus, however, is more frequent in patients with type III hyperlipoproteinemia than in control subjects [Havekes et al., 1986].

The LDL Receptor Gene

A *Pvu*II dimorphic site in the 5' of the LDL receptor gene (chromosome 19; Francke et al., [1986]) has been used as a marker to follow segregation of the FH allele in isolated families [Humphries et al., 1985; Leppert et al., 1986]. But this *Pvu*II site does not correlate with dyslipoproteinemias or FH diagnosis in association studies of unrelated individuals. One explanation in this case is that FH, due to LDL receptor deficiency, is actually an heterogeneous group of defects. Several independent molecular defects in the LDL receptor gene have indeed been shown to be the cause of FH in independent families [Tolleshaug et al., 1983; Lehrman *et al.*, 1985; Horsthemke et al., 1985].

The Insulin Gene

Insulin, involved in glucose regulation, also exerts a modulating effect on the activity of lipoprotein lipase in tissues, thereby regulating the catabolism of triglyceride-rich lipoproteins. The gene coding for human insulin has been isolated and sequenced [Bell et al., 1980b], and assigned to the short arm of chromosome 11 [Owerbach et al., 1981].

A highly polymorphic region is present in the 5' flanking sequence of the human insulin gene [Bell et al., 1980a], and is composed of 14 bp tandemly repeated DNA segments that fall into three general size classes, designated L or class 1 (0–600 bp insertion), M or class 2 (600–1,600 bp insertion), and U or class 3 (1,600–2,600 bp insertion). The insertion sequences follow classical Mendelian patterns of inheritance and are transmitted as stable genetic elements [Owerbach et al., 1982b], making them useful genetic markers for this region of chromosome 11. Presence of the U allele has been associated with atherosclerosis and macro-angiopathy in Danish populations [Owerbach et al., 1982a; Mandrup-Poulsen et al., 1984], but not among British [Jowett et al., 1984]. However, Funke et al. [1986] subsequently reported that the U allele is associated with a 3.2 times increased coronary atherosclerosis incidence in Germans.

Recommendations for Future Research

The clinical concept known as atherosclerosis represents a heterogeneous group of disorders whereby several independent molecular mechanisms lead to the same final end point – deposition of cholesterol in arterial wall cells. A comprehensive study of atherosclerosis thus involves studying all of these molecular mechanisms, some of which are known, and some others which are yet unknown. Techniques of molecular genetics represent a promising tool to reach this goal. For this reason, several groups have undertaken the task of initiating comprehensive research strategies. These consist in performing association studies on large populations of random, unrelated patients in order to unveil relationships between genetic markers and clinical manifestations of atherosclerosis. After evidencing marker-disease correlations, pedigree analyses should be performed on family members of those individuals who show positive correlations.

The converse approach consists in initating linkage studies in a few selected families. In this case, one only investigates the particular disease these isolated families suffer from. This approach is best applied to monogenic disorders of known origin, such as FH. For polygenic disorders of unknown causes (most cases of atherosclerosis), comprehensive association studies are preferable since all subtypes of the disease can be investigated at the same time.

In association studies, when a correlation between a genetic marker and a clinical variable is found, four points should be kept in mind. First of all, no cause-to-effect relationship can be established from the correlation. Because the detected DNA variation is likely not to be the defect itself but is only associated with it, the strength of the association is an inverse function of the amount of recombination between the two events, i.e. their relative physical separation. Second, the clinical manifestations under study result from multifactorial and polygenic interactive effects. Consequently, even a DNA variation which causes a molecular defect contributing to a given pathological phenotype will only be partially associated with this phenotype, because other genes also participate in its full expression. As a corollary, a weak marker-disease correlation indicates either recombination between the marker and a locus playing a major role in delineating the phenotype, or strong association with an altered gene which has low penetrance in the expression of the phenotype, or both. Third, the study of a heterogeneous group of subjects with mixed ethnic backgrounds or the phenomenon of genetic drift can lead to type I errors. Fourth, if a particular gene defect arose in multiple, independent geographic locations, the strength of the association between the defect and a marker will vary considerably. In this case, haplotype determinations will be necessary to discriminate among the various geographic origins.

References

Anderson, R. A.; Wallace, R. B.: Apoprotein A-I linked genetic polymorphisms associated with high density lipoprotein levels. Clin. Res. *33:* 15 (1985).

Assmann, G.: Lipid metabolism and atherosclerosis (Schattauer, Stuttgart 1982).

Assmann, G.; Coleman, R. T.; Funke, H.; Frossard, P. M.: Lack of association between RFLPs of the human apolipoprotein CII gene locus and clinical manifestations of atherosclerosis. Am. J. hum. Genet. *39:* suppl., abstr. 260 (1986).

Baxter, J. D.; Perloff, D.; Hsueh, W.; Biglieri, E. G.: The endocrinology of hypertension; in Felig, Baxter, Broadus, Frohmn, Endocrinology and metablolism; 2nd ed., pp. 694–788 (McGraw-Hill, New York 1987).

Bell, G. I. ; Pictet, R. L.; Rutter, W. J.: Analysis of the regions flanking the human insulin gene and sequence of an Alu family member. Nucl. Acids Res. *8:* 4091–4109 (1980a).

Bell, G. I.; Pictet, R. L.; Rutter, W. J.; Cordell, B.; Tischer, E.; Goodman, H. M.: Sequence of the human insulin gene. Nature *284:* 26–32 (1980b).

Breslow, J. L.: Human apolipoprotein molecular biology and genetic variation. A. Rev. Biochem. *54:* 699–727 (1985).

Brown, M. S.; Goldstein, J. L.: A receptor-mediated pathway for cholesterol homeostasis. Science *232:* 34–47 (1986).

Chakravarti, A.; Elbein, S. C.; Permutt, M. A.: Evidence for increased recombination near the human insulin gene: implication for disease association studies. Proc. natn. Acad. Sci. USA *83:* 1045–1049 (1986).

Ferns, G. A. A.; Galton, D. J.: Haplotypes of human apolipoprotein AI-CIII-AIV gene cluster in coronary atherosclerosis. Hum. Genet. *73:* 245–249 (1986a).

Ferns, G. A. A.; Galton, D. J.: Frequency of XbaI polymorphism in myocardial infarct survivors. Lancet *ii:* 572 (1986b).

Ferns, G. A. A.; Shelley, C. S.; Stocks, J.; Rees, A.; Paul, H.; Baralle, F.; Galton, D. J.: A DNA polymorphism of the apoprotein AII gene in hypertriglyceridaemia. Hum. Genet. *74:* 303–306 (1986).

Ferns, G. A. A.; Stocks, J.; Ritchie, C.; Galton, D. J.: Genetic polymorphisms of apolipoprotein C-III and insulin in survivors of myocardial infarction. Lancet *i:* 300–303 (1985).

Fleiss, J. L.: Statistical methods for rates and proportions (Wiley, New York 1981).

Francke, U.; Brown, M. S.; Goldstein, J. L.: Assignment of the human gene for the low density lipoprotein receptor to chromosome 19: Synteny of a receptor, a ligand, and a genetic disease. Proc. natn. Acad. Sci. USA *81:* 2826–2830 (1986).

Frossard, P. M.; Coleman, R.; Funke, H.; Assmann, G.: Molecular genetics of the human apoAI-CIII-AIV gene complex. Application to detection of susceptibility to atherosclerosis. Rhein.–Westf. Akad. Wissen. *76:* 53–62 (1987).

Frossard, P. M.; Funke, H.; Coleman, R. T.; Assmann, G.: Genetic markers for coronary atherosclerosis in the human apolipoprotein AI-CIII-AIV gene complex. Am. J. hum. Genet. *39:* suppl., abstr. 589 (1986).

Funke, H.; Frossard, P. M.; Coleman, R. T.; Assmann, G.: Genetic marker for atherosclerosis at the human insulin gene locus. Am. J. hum. Genet. *39:* suppl., abstr. 590 (1986).

Glueck, C. J.; Laskarzewski, P.; Rao, D. C.; Morrison, J. A.: Familial aggregations of coronary risk factors; in Connor, Bristow, Coronary heart disease, pp. 173–193 (Lippincott, Philadelphia 1985).

Hegele, R. A.; Breslow, J. L.: Apolipoprotein genetic variations in the assessment of atherosclerosis susceptibility. Genet. Epidemiol. *4:* 163–184 (1987).

Hegele, R. A.; Huang, L. S.; Herbert, P. N.; Blum, C. B.; Buring, J. E.; Hennekens, C. H.; Breslow, J. L.: Apolipoprotein B-gene DNA polymorphisms associated with myocardial infarction. New Engl. J. Med. *315:* 1509–1515 (1986).

Horsthemke, B.; Kessling, A. M.; Seed, M.; Wynn, V.; Williamson, R.; Humphries, S. E.: Identification of a deletion in the low density lipoprotein (LDL) receptor gene in a patient with familial hypercholesterolemia. hum. Genet. *71:* 75–78 (1985).

Humphries, S. E.; Horsthemke, B.; Seed, M.; Holm, M.; Wynn, V.; Kessling, A. M.; Donald, J. A.; Jowett, N.; Galton, D. J.; Williamson, R.: A common DNA polymorphism of the low-density lipoprotein (LDL) receptor gene and its use in diagnosis. Lancet *i:* 1003–1005 (1985).

Humphries, S. E.; Jowett, N. I.; Williams, L.; Rees, A.; Vella, M.; Kessling, A.; Myklebost, O.; Lydon, A.; Seed, M.; Galton, D. J.; Williamson, R.: A DNA polymorphism adjacent to the human apolipoprotein CII gene. Mol. Biol. Med. *1:* 463–471 (1983).

Jeffries, A. J.: DNA sequence variants in Gγ-, Aγ-, δ- and β-globin genes of man. Cell *18:* 1–10 (1979).

Jowett, N. I.; Rees, A.; Caplin, J.; Williams, L. G.; Galton, D. J.: DNA polymorphisms flanking insulin gene and atherosclerosis. Lancet *ii:* 348 (1984).

Kan, Y. W.; Dozy, A.: Polymorphism of DNA sequence adjacent to human β-globin structural gene: relationship to sickle mutation. Proc. natn. Acad. Sci. USA *75:* 5631–5635 (1978).

Karathanasis, S. K.: Apolipoprotein multigene family: Tandem organization of human apolipoprotein AI, CIII, and AIV genes. Proc. natn. Acad. Sci. USA *82:* 6374–6378 (1985).

Karathanasis, S. K.; Ferris, E.; Haddad, I. A.: DNA inversion within the apolipoproteins AI/CIII/AIV-encoding gene cluster of certain patients with premature atherosclerosis. Proc. natn. Acad. Sci. USA *84:* 7198–7202 (1987).

Karathanasis, S. K.; Norum, R. A.; Zanis, V. I.; Breslow, J. L.: An inherited polymorphism in the human apolipoprotein AI gene locus related to the development of atherosclerosis. Nature *301:* 718–720 (1983).

Kessling, A. M.; Horsthemke, B.; Humphries, S. E.: A study of DNA polymorphisms around the human apolipoprotein AI gene in hyperlipidaemic and normal individuals. Clin. Genet. *28:* 296–306 (1985).

Law, A.; Powell, L. M.; Brunt, H.; Knott, T. J.; Altman, D. G.; Rajput, J.; Wallis, S. C.; Pease, R. J.; Priestley, L. M.; Scott, J.; Miller, G. J.; Miller, N. E.: Common DNA polymorphism within coding sequences of apolipoprotein B gene associated with altered lipid levels. Lancet *i:* 1301–1303 (1986).

Lehrman, M. A.; Schneider, W. J.; Siidhof, T. C.; Brown, M. S.; Goldstein, J. L.; Russell, D. W.: Mutation in LDL receptor: Alu-Alu recombination deletes exons encoding transmembrane and cytoplasmic domains. Science *227:* 140–146 (1985).

Leppert, M. F.; Hasstedt, S. J.; Holm, T.; O'Connell, P.; Wu, L.; Ash, O.; Williams, R.; White, R.: A DNA probe for the LDL receptor gene is tightly linked to hypercholesterolemia in a pedigree with early coronary disease. Am. J. hum. Genet. *39:* 300–306 (1986).

Lim, D.; Coleman, R. T.; Assmann, G.; Frossard, P. M.: Deletion of an Alu sequence in the 5' of the apolipoprotein AI gene associated with decreased serum HDL-cholesterol levels. Am. J. hum. Genet. *39:* suppl., abstr. 621 (1986).

Mandrup-Poulsen, T.; Mortensen, S. A.; Meinertz, H.; Owerbach, D.; Johansen, K.; Sorensen, H.: DNA sequences flanking the insulin gene on chromosome 11 confer risk of atherosclerosis. Lancet *i:* 250–252 (1984).

Martin, J. B.: Genetic linkage in neurologic diseases. New Engl. J. Med. *316:* 1018–1020 (1987).

Morris, S. W.; Price, W. H.: DNA sequence polymorphisms in the apolipoprotein AI/CIII gene cluster. Lancet *ii:* 1127–1128 (1985).

Newman, W. P. III; Freedman, D. S.; Voors, A. W., et al.: Relation of serum lipoprotein levels and systolic blood pressure to early atherosclerosis: The Bogalusa Heart Study. New Engl. J. Med. *314:* 138–144 (1986).

Ordovas, J. M.; Schaeffer, E. J.; Salem, D.; Ward, R. H.; Glueck, C. J.; Vergani, C.; Wilson, P. W. F.; Karathanasis, S. K.: Apolipoprotein A-I gene polymorphism associated with premature coronary artery disease and familial hypoalphalipoproteinemia. New Engl. J. Med. *314:* 671–677 (1986).

Owerbach, D.; Bell, G. I.; Rutter, W. J.; Brown, J. A.; Shows, B.: The insulin gene is located on the short arm of chromosome 11 in humans. Diabetes *30:* 267–270 (1981).

Owerbach, D.; Billesbolle, P.; Schroll, L. M.; Johansen, K.; Poulsen, S.; Nerup, J.: Possible association between DNA sequences flanking the insulin gene and atherosclerosis. Lancet *ii:* 1291–1293 (1982a).

Owerbach, D.; Billesbolle, P.; Poulsen, S.; Nerup, J.: DNA insertion sequences near the insulin gene affect glucose regulation. Lancet *i:* 880–883 (1982b).

Rees, A.; Shoulders, C. C.; Stocks, J.; Galton, D. J.; Baralle, F. E.: DNA polymorphism adjacent to the human apoprotein A-I gene: Relation to hypertriglyceridemia. Lancet *ii:* 444–446 (1983).

Rees, A.; Stocks, J.; Sharpe, C. R.; Vella, M. A.; Shoulders, C. C.; Katz, J.; Jowett, N. I.,

Baralle, F. E.; Galton, D. J.: Deoxyribonucleic acid polymorphism in the apolipoprotein AI-CIII gene cluster. J. clin. Invest. 76: 1090–1095 (1985a).

Rees, A; Stocks, J.; Williams, L. G.; Caplin, J. L.; Jowett, N. I.; Camm, A. J.; Galton,
D. J.: DNA polymorphisms in the apolipoprotein CIII and insulin genes and atherosclerosis. Atherosclerosis 58: 269–275 (1985b).

Schulte, H.; Funke, H.; Frossard, P. M.; Coleman, R. T.; Assmann, G.: The MspI RFLP 3'
to the human apolipoprotein AII gene is neutral with respect to atherosclerosis in
Germans. Am. J. hum. Genet. 39: suppl., abstr. 293 (1986).

Scott, J.; Knott, T. J.; Priestley, L. M.; Robertson, M. E.; Mann, D. V.; Kostner, G.; Miller, G. J.; Miller, N. E.: High density lipoprotein composition is altered by a common
DNA polymorphism adjacent to apolipoprotein A-II gene in man. Lancet i: 770–773
(1985).

Southern, E. M.: Detection of specific sequences among DNA fragments separted by gel
electrophoresis. J. molec. Biol. 98: 503–517 (1975).

Suarez, B.; Cox, N.: Linkage analysis for psychiatric disorders. I. Basic concepts. Psychiat.
Dev. 3: 219–243 (1985).

Talmud, P.; Humphries, S.: DNA polymorphisms and the apolipoprotein B gene. Lancet ii:
1986 (1986).

Tolleshaug, H.; Hobgood, K. K.; Brown, M. S.; Goldstein, J. L.: The LDL receptor locus
in familial hypercholesterolemia: Multiple mutations disrupt transport and processing
of a membrane receptor. Cell 32: 941–951 (1983).

Ward, R. H.; Marshall, H. W.; Leppert, M.; Wu, L. L.; Anderson, J. L.; O'Connell, P.;
White, R. L.: Apolipoprotein genes, lipid profiles and risk of coronary artery disease
(CAD). Am. J. hum. Genet. 41: suppl., abstr. 322 (1987).

Weissman, S. M.: Molecular genetic techniques for mapping the human genome. Mol. Biol.
Med. 4: 133–143 (1987).

Williams, R.: in Hurst., update. IV. The heart, pp. 89–118 (McGraw-Hill, New York
1979).

Zannis, V. I.; Breslow, J. L.: Genetic mutations affecting human lipoprotein metabolism.
Adv. hum. Genet. 14: 125–215 (1985).

Philippe M. Frossard, PhD, California Biotechnology Inc., 2450 Bayshore Parkway,
Mountain View, CA 94043 (USA)

Lusis A J, Sparkes S R (eds): Genetic Factors in Atherosclerosis: Approaches and Model Systems. Monogr Hum Genet. Basel, Karger, 1989, vol 12, pp 125–138

Animal Models: The Watanabe Heritable Hyperlipidemic Rabbit

Brian J. Van Lenten[1]

Department of Medicine, UCLA School of Medicine, Los Angeles, Calif., USA

Introduction

Familial hypercholesterolemia (FH) is one of the most common genetic diseases among humans, characterized by a massive elevation in plasma low density lipoprotein (LDL) in its homozygous form [1]. The defect responsible for the elevated LDL levels is an allelic mutation in the gene for the membrane receptor for LDL. Consequently, there is an impaired rate of LDL catabolism, an eventual deposition of cholesterol in the arterial wall, and ultimately the development of atherosclerotic plaques. Until recently, hypercholesterolemia was studied experimentally in laboratory animals by inducing exogenous hypercholesterolemia with a high cholesterol diet. However, the transport of cholesterol among the body tissues is facilitated by lipoproteins that are of both endogenous *and* exogenous origin.

An exciting animal model has become available that closely resembles human FH. Named after the investigator who first developed the strain, the Watanabe Heritable Hyperlipidemic (WHHL) rabbit has markedly elevated plasma cholesterol levels and spontaneously develops atherosclerosis and tissue xanthoma. A great number of studies conducted since its introduction has established the WHHL rabbit as a highly suitable animal model for the study of human FH that offers the opportunity to study not only genetic abnormalities in cholesterol metabolism but also modes of intervention to prevent the development of atherosclerosis.

[1] The author wishes to thank Betty Morgan for the preparation of the manuscript and Dr. Alan Fogelman for his critical reading of the manuscript. Grant support from USPHS Grant HL30568 is also gratefully acknowledged.

Table I. Comparison of serum lipid levels with age in the WHHL rabbit and in the Japanese white rabbit

	Cholesterol mg/dl			Triglyceride mg/dl		
	1 month	6 months	12 months	1 month	6 months	12 months
WHHL rabbits (n = 154)	513 ± 129	510 ± 121	385 ± 90	571 ± 164	354 ± 88	304 ± 115
Japanese white rabbits (n = 11)	65 ± 27	41 ± 10	46 ± 16	55 ± 20	34 ± 14	49 ± 30

Adapted from Watanabe [2].

History of the WHHL Rabbit

In 1973, Watanabe discovered a single mutant rabbit that expressed elevated serum levels of cholesterol and triglyceride compared to those normally found in Japanese white rabbits. Through a controlled regimen of inbreeding [2], a strain of rabbits was produced that exhibited serum lipid values 8- to 14-fold higher than control animals (table I). Electrophoretic analysis of their serum lipoproteins revealed a broad band for β-lipoproteins and a diminished band for α-lipoproteins. In addition, these rabbits, named WHHL in 1980, were found upon necropsy to have aortic atherosclerosis and xanthoma of the digital joints.

Characterization of the WHHL LDL Receptor Defect

That this disease pattern was strikingly similar to that of human FH was further demonstrated by Tanzawa and co-workers [3, 4]. In their studies (table II), skin fibroblasts from WHHL rabbits were shown to have a higher sterol-synthesizing capacity than normal rabbit cells. This was due to a 5-fold greater activity of 3-hydroxy-3-methylglutaryl-CoA reductase (HMG-CoA reductase), the rate-limiting step in cholesterol biosynthesis [3]. In contrast, the rate of [^{14}C]oleate incorporation into cholesteryl esters in the mutant cells was 9-fold lower than in normal cells. Neither the levels of HMG-CoA reductase activity nor of the cholesterol-esterification activity of WHHL fibroblasts could be altered by either the addition of LDL to or its removal from the culture medium, indicating an absence of functional LDL receptors on WHHL fibroblasts.

Table II. Activities of LDL receptor, HMG-CoA reductase and cholesterol esterification in skin fibroblasts from normal and WHHL rabbits [3]

Rabbit strain	Number of cell strains	^{125}I-LDL, ng·2 h^{-1}·mg^{-1}		HMG-CoA reductase[a] pmol·min^{-1}·mg^{-1}	Cholesterol esterification[a] pmol·h^{-1}·mg^{-1}
		high-affinity binding	high-affinity degradation		
Normal	3	342 ± 22	796 ± 117	4.7 ± 0.9	533 ± 38
WHHL	6	7.6 ± 5.8	10.6 ± 3.8	31.3 ± 6.5	54.5 ± 2.9

[a] The enzyme activites were assayed after 3 days of incubation in medium containing 10% fetal calf serum.

Indeed, experiments studying the binding and degradation of ^{125}I-labeled LDL demonstrated that WHHL fibroblasts possessed only 6% of the normal rabbit's level of LDL receptor activity, whereas cell lines from heterozygous WHHL rabbits had about the same LDL receptor activities as normal cells [4]. These results suggested that the primary defect in WHHL rabbit fibroblasts is a deficiency of LDL receptors and that the mode of inheritance of LDL receptor deficiency is recessive. Also consistent with an LDL receptor deficiency was the observation that after injection of ^{125}I-LDL there was a reduced rate of clearance of the ^{125}I-LDL from the blood of WHHL rabbits compared to normal controls [3].

How widespread is the lack of LDL receptors in WHHL rabbits? Kita et al. [5] reasoned that if tissues such as the liver and adrenal gland, organs important in cholesterol metabolism, normally possessed LDL receptor activity, then these receptors should be genetically similar to the well-characterized fibroblast receptor. They found that a Ca^{2+}-dependent binding site, similar to the fibroblast LDL receptor, was indeed severely deficient or absent in the livers and adrenals of the WHHL rabbits studied. They concluded therefore that these receptor deficiencies doubtless contributed to the hypercholesterolemia of the animals.

LDL Receptor-Independent Processes

The presence of LDL receptors on cells provides an efficient means for sequestering and internalizing extracellular cholesterol, reducing the energy requirements necessary for the cell to synthesize its own cholesterol. Normally, as much as 70% of the total plasma LDL-cholesterol can be removed by the LDL receptors of the rabbit liver [6]. In the WHHL rabbit, however, essentially all of the LDL appeared to be catabolized by a process independent of the LDL receptor [7]. Nevertheless, the receptor-independent FCR in WHHL rabbits was in fact similar to that in normal rabbits (table III). Because of the very high plasma LDL concentrations in the WHHL rabbit, catabolism of LDL by receptor-independent processes is greater than that catabolized by the LDL receptor in the normal rabbit [8]. In the receptor-deficient WHHL rabbit, most of the LDL-cholesterol uptake could be accounted for by extrahepatic tissues via much slower receptor-independent processes [6]. One can envision how in the WHHL rabbit, as in the FH patient, a loss of LDL receptor activity could lead to not only an increased plasma LDL-cholesterol concentration but also a change in the organ distribution for LDL degradation. Consequently, the relative amount of LDL-cholesterol taken up by a tissue (such as the artery wall) could be much greater than normal.

Lipoprotein Removal and Cholesterol Balance in the WHHL Rabbit

In vitro the LDL receptor has been shown to bind lipoproteins containing both apolipoprotein (apo) B and apo E, hence its alternative name; the apo B, E receptor [9]. These receptors can therefore bind apo B-rich lipoproteins such as LDL as well as the apo E-rich lipoproteins such as β-migrating very low density lipoprotein (β-VLDL) and high density lipoprotein from cholesterol-fed animals (apo E-HDL$_c$). The functional absence of the LDL receptor in the WHHL rabbit provided investigators with the opportunity to study its role in the clearance of other lipoproteins from the plasma. After a meal, chylomicrons produced by the intestine are released into the blood and enzymatically converted into remnant particles that are removed by the liver [10]. The liver in turn synthesizes and secretes VLDL that are also converted to intermediate density lipoproteins (IDL) and eventually to LDL. Kita et al. [11] have demonstrated that when both normal and WHHL rabbits were given intravenous injections of labeled rat chylomicrons, all radiolabeled components were removed from the plasma at comparable rates in both strains (fig. 1). Additionally, comparable amounts of radioactivity accumulated in the livers of both normal and WHHL rabbits. Isolated liver membranes from WHHL rabbits, however, were deficient in the binding of ^{125}I-labeled chylomicron remnants, suggesting that in part, chylomicron remnants bind to the LDL receptor. So it appeared that WHHL rabbits possess a normal chylomicron remnant uptake mechanism that allows them

Table III. Kinetic parameters for LDL and methyl-LDL turnover in normal and WHHL rabbits after Bilheimer et al. [7]

Rabbit	Total plasma apo-LDL pool, mg	FCR[a], pools per day			Rate of catabolism of apo-LDL, mg/kg/day		
		total (a)	receptor-independent (b)	receptor-mediated (a–b)	total (c)	receptor-independent (d)	receptor-mediated (c–d)
Normal							
A-1	15	1.54	0.27	1.27	8.7	1.5	7.2
A-2	18	2.00	0.55	1.45	13.1	3.6	9.5
B-1	6	1.80	0.44	1.36	3.5	0.8	2.7
B-2	5	1.44	0.95	0.49	3.2	2.1	1.1
C-1	18	1.40	0.64	0.76	6.0	2.7	3.3
C-2	19	1.69	0.37	1.32	8.6	1.9	6.7
Mean ± SEM	14 ± 3	1.65 ± 0.09	0.54 ± 0.10	1.11 ± 0.16	7.2 ± 1.5	2.1 ± 0.4	5.1 ± 1.3
WHHL[b]							
A-3	201	0.36	0.28	0.08	30.0	23.3	6.7
A-4	318	0.38	0.50	(−0.12)	44.1	58.1	(−14)
B-3	249	0.49	0.37	0.12	44.9	34.1	10.8
B-4	274	0.51	0.46	0.05	45.5	41.0	4.5
C-3	184	0.62	0.76	(−0.14)	38.6	47.3	(−8.7)
Mean ± SEM	245 ± 24	0.47 ± 0.05	0.47 ± 0.08	0.00 ± 0.05	40.6 ± 1.9	40.8 ± 5.9	(−0.14) ± 4.8
p values[c]	0.005	0.005	S	0.005	0.005	0.005	NS

[a] FCR of apo-LDL or methyl-apo-LDL from the intravascular space. Apo-LDL = LDL-associated protein.

[b] The intravenous catheter in animal C-4 became inoperative before sufficient data for the turnover study were collected.

[c] p values refer to statistical comparisons between the normal and WHHL groups for each of the parameters listed; NS = not significant.

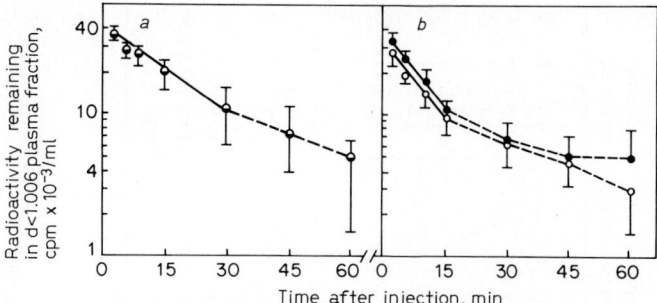

Fig. 1. Disappearance of [¹⁴C]triglycerides (●) and [³H]cholesteryl esters (○) from the plasma fraction of d <1.006 g/ml of 4 normal rabbits *(a)* and 4 WHHL rabbits *(b)* after intravenous injection of double-labeled rat lymph chylomicrons. After Kita et al. [11].

to remove chylomicrons from the blood despite a lack of significant LDL receptor activity. Moreover, the uptake of chylomicron remnants by the liver may be mediated by a binding site that recognizes a component of the particle other than apo B, such as apo E. In contrast, radiolabeled VLDL and IDL were cleared from the plasma of WHHL rabbits much more slowly than from the plasma of normal rabbits (table IV) [12]. As a consequence of the impaired clearance rates, there was an increased conversion of VLDL to LDL in the plasma. One can see how the deficiency of LDL receptors in the WHHL rabbit, as well as in the FH patients, could lead to concomitant overproduction *and* a delayed clearance of LDL producing their markedly elevated plasma levels.

Pathology of the Atherosclerosis in WHHL Rabbits

Atherosclerosis in humans is often characterized by fibrous plaques that have accumulated lipid and contain connective tissue caps of collagen and elastic fibers that cover the lipids [13]. This intimal accumulation of lipid is associated with an increase in lipid-containing cells of smooth muscle cell and macrophage origin, called foam cells [14]. Buja et al. [15] conducted a detailed examination of the atherosclerotic lesions of WHHL rabbits, extending the initial observations of Watanabe [2]. Atherosclerosis in WHHL rabbits was characterized by the evolution of progressive intimal lesions. These lesions included not only fatty streaks and medial lipid deposition but also raised foam cells and plaques. In the coronaries

Table IV. Fate of ^{125}I-apo B after intravenous injection of ^{125}I-VLDL into normal and WHHL rabbits

Exp.	VLDL donor[a]	Lipoprotein fraction	Percent of injected dose	
			normal	WHHL
A	WHHL	total	9.4	52
		VLDL	6.7	22
		IDL	1.8	22
		LDL	0.9	8.0
B	normal	total	7.8	30
		VLDL	2.3	10
		IDL	2.5	13
		LDL	3.0	7.1
C	normal	total	27	62
		VLDL	3.0	4.6
		IDL	11	22
		LDL	13	35

Adapted from Kita et al. [12].

Experiments: A, recipient normal (n = 3) and WHHL (n = 4) rabbits (fed) were injected with 2.9×10^8 cpm of ^{125}I-VLDL prepared from 24-hour fasted WHHL rabbits; B, recipient normal (n = 2) and WHHL (n = 2) rabbits were fasted 16 h prior to the injection of 2.3×10^8 cpm of ^{125}I-VLDL prepared from 24-hour fasted normal rabbits; C, recipient normal (n = 2) and WHHL (n = 2) rabbits (all female and fed) were injected intravenously with 2.6×10^8 cpm of ^{125}I-VLDL prepared from 24-hour fasted normal rabbits. Blood was obtained 6 h (experiments A and C) or 7 h (experiment B) after injection of ^{125}I-VLDL, and the plasma was subjected to sequential ultracentrifugation. The ^{125}I-apo B content of each lipoprotein fraction was measured by isopropanol precipitation and is expressed as a percentage of the ^{125}I-apo B present in the original ^{125}I-VLDL. The values were not corrected for procedural losses, which averages 5–10% for each lipoprotein fraction. Each value shown is the average of data obtained from the indicated number of animals.
[a] All fasted 24 h.

and the aorta the lesions were found to be rich in both unesterified and esterified cholesterol. Lipid accumulation in smooth muscle cells appeared to be a predominant feature of these arterial intimal lesions. The accumulation of lipid in macrophages also appeared to contribute to the growth of the plaque. However, only mild lipid accumulation was found in other tissues, such as the spleen, lymph nodes, and liver. In these tissues the lipid was usually found within macrophages. This pattern contrasted with that seen in the normal rabbit fed a high-cholesterol diet [16]. These animals develop arterial lesions of a pure foam cell type, primarily of macrophage origin. Moreover, there was a much wider dis-

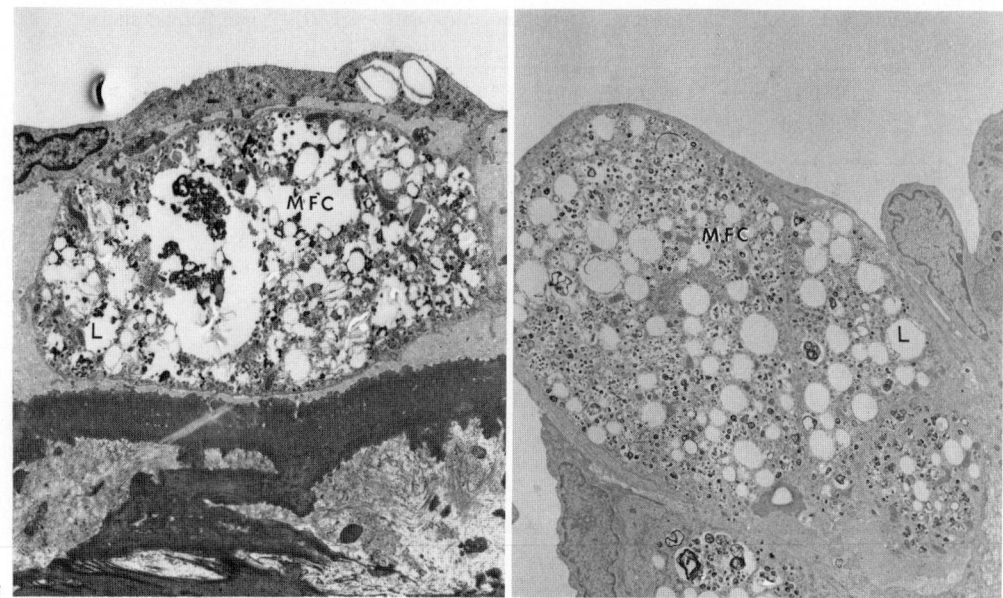

Fig. 2. Transmission electron micrographs of subendothelial macrophage-derived foam cells in fatty streaks of WHHL rabbits. *a* Thoracic aorta of a 43-day-old WHHL rabbit. ×10,000. *b* Aortic arch from a 5-month-old WHHL rabbit. ×800. L = lipid droplets; MFC = macrophage-derived foam cell. Lead citrate-uranyl acetate counterstain. Micrographs kindly provided by Dr. M. Rosenfeld.

tribution of lipid accumulation in other tissues of the cholesterol-fed rabbit.

Rosenfeld and co-workers [17–19] carried out similar studies, but utilized cell-specific monoclonal antibodies to characterize the lesions of WHHL rabbits and cholesterol-fed rabbit. In their studies the chronology of cellular interactions that occurred during the development of the fatty streak and in the expansion of these early lesions into mature lesions was similar for both types of rabbits. This included the appearance of macrophage-derived foam cells in the early fatty streak [18]. Smooth muscle cells were observed simultaneously with macrophages in the intima of young rabbits. As the lesions expanded and matured, the authors observed alternating layers of macrophages and smooth muscle cells with a hypertrophy of the macrophage-derived foam cells beneath the endothelium [19]. Endothelial retraction could also be observed, exposing the subendothelial foam cells. Eventually thrombus formation on the surface of the exposed macrophage-derived foam cells could be seen. This lesion expansion involved the continued attachment and migration of monocytes into the intima that would eventually accumulate lipids and grow in size (fig. 2). Thus, many morphological similarities exist between human atherosclerosis and the disease in WHHL rabbits. However, one sees from this model that the contribution of the LDL receptor to the development of arterial lesions can be quite small.

Cholesterol Accumulation in the Absence of LDL Receptors

The above results raise some important questions about the accumulation of cholesteryl esters in the lesions of WHHL rabbits. Most of the cholesterol carried in the blood of WHHL rabbits is in the form of apo B-containing lipoproteins, primarily LDL [20]. WHHL cells lack significant LDL receptor activity but can clear chylomicron remnants quite effectively [3–5, 11]. Significantly, the macrophages of WHHL rabbits, although lacking specific LDL receptor activity, can take up and degrade apo E-rich chylomicron remnants and β-VLDL [21]. This could provide a mechanism for the accumulation of cholesteryl esters in the macrophage-derived foam cells of the lesions of WHHL rabbits in spite of an inability to take up and degrade LDL by receptor-mediated processes. Further evidence for this hypothesis has been provided by Kita et al. [22]. When they incubated the VLDL from WHHL or normal Japanese rabbits with mouse peritoneal macrophages, the VLDL from WHHL rabbits stimulated cholesteryl ester synthesis 124-fold more than did VLDL from control animals. WHHL-VLDL and β-VLDL competed for the same cellular site on the macrophage but LDL did not. Thus, the ability of cholesterol-rich VLDL from WHHL rabbits to produce cholesteryl ester storage in macrophages appeared to be dependent on a process other than the LDL receptor.

The question of whether or not a distinct receptor exists for cholesteryl ester-rich lipoproteins such as β-VLDL has been recently addressed by a number of investigators. Koo et al. [23] showed that an unusual apo B, E (LDL) receptor was responsible for the binding of β-VLDL to mouse peritoneal macrophages. This receptor binds LDL poorly but binds apo E-containing lipoproteins normally and is resistant to down-regulation by cholesterol. Antibody to the classical LDL receptor has the ability to block the binding of β-VLDL as well as LDL to murine macrophages [23, 24]. Work from this laboratory [25] has shown that although aortic endothelial cells from a WHHL rabbit were able to take up and degrade β-VLDL similarly to normal rabbit cells, the uptake process was inhibited by antibody made against the LDL receptor. It would appear then that the receptor responsible for the uptake of β-VLDL by WHHL cells is at least immunologically related to the classical LDL receptor.

One cannot discount the role that another receptor may play in the development of atherosclerosis in WHHL rabbits. The scavenger receptor has been shown to recognize modified lipoproteins such as acetylated LDL, the uptake of which can lead to cholesteryl ester accumulation [26]. If modification of normal LDL does occur in vivo, as has been suggested [26, 27], the scavenger receptor could effectively remove these altered LDL, resulting in the accumulation of cholesteryl esters and foam cell production.

Molecular Biology of the Receptor Defect in WHHL Rabbits

The LDL receptor is normally synthesized as a precursor having an apparent molecular weight of 120 kd [28] and a large number of immature carbodydrate chains. After processing of the carbohydrates, the mature receptor increases in size to 160 kd. Once transferred to the surface of the cell, the LDL receptor can bind the apo B- and apo E-containing lipoproteins, such as LDL and β-VLDL. Yamamoto et al. [29] have shown from analysis of the sequence of the LDL receptor that a cysteine-rich 40-amino acid repeat sequence exists at the NH_2-terminal end of the receptor. These repeat sequences appear to be important in the binding domain of the receptor and may interact with lipoproteins via the arginine and lysine residues of apo B and apo E.

Information on the nature of the defect in LDL receptor expression of WHHL rabbits has recently become available. These mutants have an allele that produces a receptor of normal molecular weight but that is transported to the cell surface at one-tenth the normal rate [30]. Although the receptor contains its normal complement of carbohydrates, the sugar chains are not processed normally by the Golgi complex. As a result, few receptors actually are transported to the cell surface. The apparent cause of this defect was obtained through DNA sequence analysis [31]. Twelve base pairs normally present in the region between cysteines 113 and 122 of the LDL receptor cDNA from rabbit were absent in the WHHL cDNA. This deletion resulted in a loss of 4 amino acids of the ligand-binding domain of the receptor. Ligand-blotting experiments were also conducted using ^{125}I-labelled β-VLDL incubated with solubilized proteins from normal and WHHL adrenal gland membranes. Interestingly, the *precursor* form of the LDL receptor found in WHHL membranes effectively bound the ^{125}I-β-VLDL. These data indicate that in the WHHL rabbit, the immature receptors that are synthesized and transported to the cell surface, although unable to bind LDL, appear to bind apo E-rich lipoproteins normally. These results could explain, in part, the relatively normal *whole-body* tissue cholesterol balance in the WHHL rabbit.

Chemotherapeutic Considerations

Conventionally, treatment of hyperlipidemia has involved drug therapy focusing on lowering of the LDL in the plasma. Compounds such as mevinolin and colestipol, lower LDL levels by increasing LDL receptor activity and hence the removal of LDL from the plasma [32]. However, in the case of FH homozygous patients who lack LDL receptors, this mode of therapy is not efficacious. Experiments in WHHL rabbits using the compound probucol could provide an attractive alternative to the drugs previously mentioned [33]. Probucol administration to WHHL rabbits in a 1% w/w chow diet lowered LDL-cholesterol levels 36%. It appeared initially from turnover studies measuring the FCR of ^{125}I-LDL prepared from untreated WHHL rabbits that the lipid-lowering effect of probucol was not due to a change in the actual FCR of LDL. However, if WHHL rabbits were first treated with probucol and then their LDL were labeled and injected back into them, the FCR increased 50% compared to 'untreated' LDL. Analysis of the LDL after probucol treatment revealed an overall decrease in its cholesterol content. Additionally, the uptake and degradation of this LDL by cultured skin fibroblasts from both WHHL and normal rabbits was in fact higher than that from untreated donors. Subsequent studies, however, from the same laboratory showed that the addition of 5 μM probucol inhibited oxidation of LDL in vitro, actually decreasing the uptake of the LDL by murine macrophages [34]. To what extent LDL is modified by probucol itself and how that modification can influence its metabolic fate remain to be determined.

Recently, probucol has been shown to be effective in preventing the progression of atherosclerosis in WHHL rabbits in vivo without an alteration in plasma cholesterol levels [35]. In vitro, probucol was shown to inhibit the oxidation of LDL by endothelial cells as well as by chemical agents such as Cu^{2+}. This could explain its inhibitory effect on lesion development by reducing the uptake of modified LDL via the scavenger receptor on the macrophage.

Conclusions

The discovery of the WHHL rabbit has provided the field of atherosclerosis research with a unique animal model. Its similarity to the human FH patient makes it highly desirable for studying the inborn errors of cholesterol metabolism in an organism that lacks functional LDL receptors. The number of laboratories utilizing the WHHL rabbit for atherosclerosis research is growing, and the results from this research are helping to elucidate the mechanisms underlying the development of atherosclerosis.

References

1   Goldstein, J. L.; Brown, M. S.: The LDL receptor locus and the genetics of familial hypercholesterolemia. Annu. Rev. Genet. *13:* 259–289 (1979).
2   Watanabe, Y.: Serial inbreeding of rabbits with hereditary hyperlipidemia (WHHL rabbit). Atherosclerosis *36:* 261–268 (1980).
3   Tanzawa, K.; Shimada, Y.; Kuroda, M.; Tsujita, Y.; Mamoru, A.; Watanabe, H.: WHHL rabbit: A low density lipoprotein receptor-deficient animal model for familial hypercholesterolemia. FEBS Lett. *118:* 81–84 (1980).
4   Shimada, Y.; Tanzawa, K.; Kuroda, M.; Tsujita, Y.; Mamoru, A.; Watanabe, Y.: Biochemical characterization of skin fibroblasts derived from WHHL rabbit, a notable animal model for familial hypercholesterolemia. Eur. J. Biochem. *118:* 557–564 (1981).
5   Kita, T.; Brown, M. S.; Watanabe, Y.; Goldstein, J. L.: Deficiency of low density lipoprotein receptors in liver and adrenal gland of the WHHL rabbit, an animal model of familial hypercholesterolemia. Proc. natn. Acad. Sci. USA *78:* 2268–2272 (1981).
6   Spady, D. K.; Huettinger, M.; Bilheimer, D. W.; Dietschy, J. M.: Role of receptor-independent low density lipoprotein transport in the maintenance of tissue cholesterol balance in the normal and WHHL rabbit. J. Lipid Res. *28:* 32–41 (1987).
7   Bilheimer, D. W.; Watanabe, Y.; Kita, T.: Impaired receptor-mediated catabolism of low density lipoprotein in the WHHL rabbit, an animal model of familial hypercholesterolemia. Proc. natn. Acad. Sci. USA *79:* 3305–3309 (1982).
8   Pittman, R. C.; Carew, T. E.; Attie, A. D.; Witztum, J. L.; Watanabe, Y.; Steinberg, D.: Receptor-dependent and receptor-independent degradation of low density lipoprotein in normal rabbits and in receptor-deficient mutant rabbits. J. biol. Chem. *257:* 7994–8000 (1982).
9   Brown, M. S.; Goldstein, J. L.: Receptor-mediated control of cholesterol metabolism. Science *191:* 150–154 (1976).
10 Green, P. H. R.; Glickman, R. M.: Intestinal lipoprotein metabolism. J. Lipid Res. *22:* 1153–1173 (1981).
11 Kita, T.; Goldstein, J. L.; Brown, M. S.; Watanabe, Y.; Hornick, C. A.; Havel, R. J.: Hepatic uptake of chylomicron remnants in WHHL rabbits: a mechanism genetically distinct from the low density lipoprotein receptor. Proc. natn. Acad. Sci. USA *79:* 3623–3627 (1982).

12 Kita, T.; Brown, M. S.; Bilheimer, D. W.; Goldstein, J..: Delayed clearance of very low density and intermediate density lipoproteins with enhanced conversion to low density lipoprotein in WHHL rabbits. Proc. natn. Acad. Sci. USA 79: 5693–5697 (1982).

13 Duff, G.; McMillan, G.: Pathology of atherosclerosis. Am. J. Med. 11: 92–108 (1951).

14 Ross, R.: Atherosclerosis: a problem of the biology of arterial wall cells and their interactions with blood components. Arteriosclerosis 1: 293–311 (1981).

15 Buja, L. M.; Kita, T.; Goldstein, J. L.; Watanabe, Y.; Brown, M. S.: Cellular pathology of progressive atherosclerosis in the WHHL rabbit. An animal model of familial hypercholesterolemia. Arteriosclerosis 3: 87–101 (1983).

16 Scott, R. F.; Daoud, A. S.; Florentin, R. A.: Animal models in atherosclerosis; in Wissler, Geer, The pathogenesis of atherosclerosis, pp. 120–146 (Williams & Wilkins, Baltimore 1971).

17 Tsukada, T.; Rosenfeld, M.; Ross, R.; Gown, A. M.: Immunocytochemical analysis of cellular components in atherosclerotic lesions. Use of monoclonal antibodies with the Watanabe and fat-fed rabbit. Arteriosclerosis 6: 601–613 (1986).

18 Rosenfeld, M. E.; Tsukada, T.; Gown, A. M.; Ross, R.: Fatty streak initiation in Watanabe heritable hyperlipemic and comparably hypercholesterolemic fat-fed rabbits. Arteriosclerosis 7: 9–23 (1987).

19 Rosenfeld, M. E.; Tsukada, T.; Chait, A; Bierman, E. L.; Gown, A. M.; Ross, R.: Fatty streak expansion and maturation in Watanabe heritable hyperlipemic and comparably hypercholesterolemic fat-fed rabbits. Arteriosclerosis 7: 24–34 (1987).

20 Havel, R. J.; Kita, T.; Kotite, L.; Kane, J. P.; Hamilton, R. L.; Goldstein, J. L.; Brown, M. S.: Concentration and composition of lipoproteins in blood plasma of the WHHL rabbit. Arteriosclerosis 2: 467–474 (1982).

21 Van Lenten, B. J.; Fogelman, A. M.; Jackson, R. L.; Shapiro, S.; Haberland, M. E.; Edwards, P. A.: Receptor-mediated uptake of remnant lipoproteins by cholesterol-loaded human monocyte-macrophages. J. biol. Chem. 26: 8783–8788 (1985).

22 Kita, T.; Yokode, M.; Watanabe, Y.; Narumiya, S.; Kawai, C.: Stimulation of cholesteryl ester synthesis in mouse peritoneal macrophages by cholesterol-rich very low density lipoproteins from the Watanabe heritable hyperlipidemic rabbit: an animal model of familial hypercholesterolemia. J. clin. Invest. 77: 1460–1465 (1986).

23 Koo, C.; Wernette-Hammond, M. E.; Innerarity, T. L.: Uptake of canine β-very low density lipoproteins by mouse peritoneal macrophages is mediated by a low density lipoprotein receptor. J. biol. Chem. 261: 11194–11201 (1986).

24 Ellsworth, J. L.; Kraemer, F. B.; Cooper, A. D.: Transport of β-very low density lipoproteins and chylomicron remnants by macrophages is mediated by the low density lipoprotein receptor pathway. J. biol. Chem. 262: 2316–2325 (1987).

25 Fogelman, A. M.; Berliner, J. A.; Van Lenten, B. J.; Navab, M.; Territo, M.: Lipoprotein receptors and endothelial cells. Semin. Thromb. Hemost. (in press).

26 Goldstein, J. L.; Ho, Y. K.; Basu, S. K.; Brown, M. S.: Binding site on macrophages that mediates uptake and degradation of acetylated low density lipoprotein, producing massive cholesterol deposition. Proc. natn. Acad. Sci. USA 76: 333–337 (1979).

27 Fogelman, A. M.; Shechter, I.; Seager, J.; Hokom, M.; Child, J. S.; Edwards, P. A.: Malondialdehyde alteration of low density lipoproteins leads to cholesteryl ester accumulation in human monocyte-macrophages. Proc. natn. Acad. Sci. USA 77: 2214–2218 (1980).

28 Goldstein, J. L.; Brown, M. S.; Anderson, R. G.; Russell, D. W.; Schneider, W. J.: Receptor-mediated endocytosis: concepts emerging from the LDL receptor system. Annu. Rev. cell. Biol. 1: 1–39 (1985).

29   Yamamoto, T.; Davis, C. G.; Brown, M. S.; Schneider, W. J.; Casey, M. L.; Gold-
     stein, J. L.; Russell, D. W.: The human LDL receptor: a cysteine-rich protein with
     multiple Alu sequences in its mRNA. Cell *39:* 27–38 (1984).
30   Schneider, W. J.; Brown, M. S.; Goldstein, J. L.: Kinetic defects in the processing of
     the low density lipoprotein receptor in fibroblasts from WHHL rabbits and a family
     with familial hypercholesterolemia. Mol. Biol. Med. *1:* 353–367 (1983).
31   Yamamoto, T.; Bishop, R. W.; Brown, M. S.; Goldstein, J. L.; Russell, D. W.: Dele-
     tion in cysteine-rich region of LDL receptor impedes transport to cell surface in
     WHHL rabbit. Science *232:* 1230–1237 (1986).
32   Bilheimer, D. W.; Grundy, S. M.; Brown, M. S.; Goldstein, J. L.: Mevinolin and col-
     estipol stimulate receptor-mediated clearance of low density lipoprotein from plasma
     in familial hypercholesterolemia heterozygotes. Proc. natn. Acad. Sci USA *80:* 4124–
     4128 (1983).
33   Naruszewicz, M.; Carew, T. E.; Pittman, R. C.; Witztum, J. L.; Steinberg, D.: A
     novel mechanism by which probucol lowers low density lipoprotein levels demon-
     strated in the LDL receptor-deficient rabbit. J. Lipid Res. *25:* 1206–1213 (1984).
34   Parthasarathy, S.; Young, S. G.; Witztum, J. L.; Pittman, R. C.; Steinberg, D.: Prob-
     ucol inhibits oxidative modification of low density lipoprotein. J. clin. Invest. *77:* 641–
     644 (1986).
35   Kita, T.; Nagano, Y.; Yokode, M.; Ishii, K.; Kume, N.; Ooshima, A.; Yoshida, H.;
     Kawai, C.: Probucol prevents the progression of atherosclerosis in Watanabe heritable
     hyperlipidemic rabbit, and animal model for familial hypercholesterolemia. Proc.
     natn. Acad. Sci. USA *84:* 5928–5931 (1987).

Brian J. Van Lenten, PhD, Department of Medicine, UCLA School of Medicine,
Los Angeles, CA 90024 (USA)

Lusis A J, Sparkes S R (eds): Genetic Factors in Atherosclerosis: Approaches and
Model Systems. Monogr Hum Genet. Basel, Karger, 1989, vol 12, pp 139–169

Animal Models: The Pig

Jan Rapacz, Judith Hasler-Rapacz

Department of Genetics and Department of Meat and Animal Science,
University of Wisconsin, Madison, Wisc., USA

Atherosclerotic disease in man, which includes coronary, renal, cerebral and peripheral vascular diseases, has reached an alarmingly high incidence in the affluent nations. This disease is the major cause of disability
and accounts for approximately half of all deaths in the United States
every year. Atherosclerosis is visualized as a complex, multifactorial, and
dynamic polypathogenic process involving vascular tissues affected by dis-
turbances in metabolic processes leading to arterial lesions, which develop into a severe disease. Owing to these complexities the disease process is still poorly understood. Although atherosclerotic lesions can be
induced experimentally in animals by a single factor (for example, high
cholesterol in the diet), in humans they begin in childhood, spontaneously and undetected, and develop irregularly over a long life span into a
severe atherosclerotic occlusive disease in older adults.

This insidious process represents a number of distinctly different clinical and morphopathologic patterns having various rates of progression,
essentially without showing clinical effects, until the artery is nearly
occluded. This impedes blood circulation and triggers a sudden appearance of symptoms in the form of a heart attack, sudden death, stroke,
angina or claudication. In addition to the common difficulties in detecting
the stage of the preclinical lesions, there are individual inherent differences and innumerable exogenous factors. These include different life
habits and diets which contain highly variable kinds and contents of fats
and energy, making investigations of the atherosclerotic process in
humans difficult or infeasible. Therefore, reliable experimental animal
models are desirable in an attempt to further advance and elucidate the
nature of the atherosclerotic process in humans.

Experimental models provide opportunities to explore underlying
mechanisms of the disease process and to identify both exogenous and

endogenous factors which can modify metabolic processes. In addition, models enable studies of the alleged atherogenicity of numerous variables and associated events under selected experimental conditions, and allow for comparisons with the clinically observed disease in humans. A variety of animal models have been developed during the past four decades in response to these needs. Owing primarily, but not exclusively to inherent species differences, no animal model can duplicate the human disease.

Although, around 1500, Leonardo da Vinci used swine in medical studies demonstrating the movement of the heart during the cardiac cycle [56], the pig has seldom been used in cardiovascular research until 1954 when Gottlieb and Lalich [29] reported on naturally occurring atherosclerosis in swine. At that time domestic swine were represented by a large number of breeds worldwide (a rough estimate by the author is over 200 breeds), and provided a vast array of genetic variation between and within breeds. This variation offered useful large animal material for research in atherosclerosis. Similarities in size, physiology, cardiovascular structure and coronary artery distribution with humans as well as an omnivorous nature [6, 19] made swine a suitable species to explore further its usefulness as a model.

Studies were undertaken by a number of investigators who focused their efforts primarily on three objectives: (1) characterization of atherosclerosis by determining grossly and histologically the prevalence, arterial distribution and severity of spontaneous atherosclerosis; (2) comparison of swine and human spontaneous atherosclerosis, and (3) exploration of the experimental induction and acceleration of the atherosclerotic process by manipulating dietary components.

Results of studies during the first 15 years showed that swine spontaneous atherosclerosis has a remarkable resemblance to the early features of the human disease, and that the disease can be accelerated by feeding diets high in saturated fat and cholesterol. Consequently, the pig became a very attractive animal model to study atherosclerosis during the last two decades. Studies were continued to determine the effects of different endogenous and exogenous factors, mechanisms and time required for the development of advanced atherosclerosis; however, the principal variable investigated was the level and type of dietary fat and cholesterol.

Data from studies of the effects of lipids on the development of accelerated atherosclerosis also showed, in over 50% of swine, a significant increase in plasma cholesterol and low density lipoproteins (LDL), the principal carrier of cholesterol. This confirmed earlier implications of LDL in atherosclerosis [27], and led to the studies of swine lipoproteins, their metabolism, alterations and involvement in atherogenesis. Investigations were expanded into studies of morphology, changes and composition of the arterial wall and lesions, identification of molecular and cellular components, cellular interactions and sequences of events occurring during the development of accelerated atherosclerosis.

Studies using swine models during the past three decades have contributed to a better understanding of human atherosclerosis. It is beyond the limits of this article to review all of these reports. The reader is directed to more detailed reviews, each covering studies of a limited period of time or area of research [6, 11, 15, 40, 53, 57, 62, 63, 65, 92, 98, 109]. The first part of this chapter is intended to present a brief summary, or cite studies which characterized swine spontaneous and experimentally induced atherosclerosis, its resemblance to the human disease, and identify atherogenic variables, as well as characteristic alterations in plasma constituents and in the morphology of the arterial wall. In addition, an attempt is made to consider individual variations among experimental swine, which were observed in a number of these studies, but rather seldom commented on their possible nature.

The second part is intended to present a summary of advances made in studies of genetic aspects of lipoprotein diversity and its possible association with atherosclerosis. Immunogenetic studies, carried out during the last two decades in this laboratory, provide insight into genetic diversity in protein constituents of the lipoproteins known as apolipoproteins (apo). A number of apolipoprotein variants (allotypes, epitopes or apoliprotein markers) have been identified. These correspond to gene mutations and represent genetic polymorphism of apolipoproteins.

The advances in knowledge of genetic polymorphism of lipoproteins in swine led to the identification of a new genetic (endogenous, or spontaneous) form of hyperlipoproteinemia and hypercholesterolemia associated with accelerated atherosclerosis. Unlike individuals with familial hypercholesterolemia [5] and WHHL rabbits [108], the affected swine have normal LDL receptor activity [84]. In addition, pigs homozygous for some of the LDL mutations (apo-B) exhibit distinct cholesterol concentrations and confer either enhanced susceptibility or resistance to diet-induced cholesterolemia.

Spontaneous Atherosclerosis in Swine

Lack of investigations on swine in comparative atherosclerosis led Gottlieb and Lalich [29] to examine grossly and microscopically lesions of spontaneous atherosclerosis in 1,775 pig aortas, collected at slaughter-houses from animals varying in age from 4 months to over 3 years. Following their findings, that in swine lesions of the aorta are common and that the incidence increases with age, several other studies were carried out to further expand investigations on spontaneous atherosclerosis and its resemblance to humans [20, 25, 45, 59, 88, 97, 102]. In addition to these reports, spontaneously developed lesions in swine were evaluated in other studies.

The above studies, which dealt exclusively with spontaneous athero-sclerosis, included 2,269 swine of both sexes, in age from birth to 14 years, representing from one to nine, and likely many more [104], breeds and crosses per study. All the animals were fed a basic low-fat ration with the fat content less than 5% [19, 20, 25, 45, 88, 97, 102] except for one study [59] in which 50 pigs were fed cooked garbage containing as much as 12% fat of unknown origin. Together, these investigations covered the entire life span of swine providing an opportunity for studying all stages of atherosclerotic changes.

Morphological descriptions of spontaneous atherosclerosis in these reports were based on gross and light microscopy [19, 20, 25, 45, 88, 97, 102] and on electron microscopy [19, 20] examinations. The investigations concentrated on the aorta and its main branches, with special attention to the coronary arteries. The first changes were usually observed at 6 months of age, occasionally in 4-month-old piglets [25], and even in a 2-month-old animal [19]. Early lesions appeared most frequently adjacent to the orifices as oval yellowish [25], or grayish [20] spots or streaks [97] projecting slightly above the intimal surface. They appeared similar to human lesions when stained with Sudan IV. The coronary arteries of a 2-month-old piglet resembled those of the human newborn, and even at this age the intima of the proximal part of the arteries was thickened by protrusion of muscle cells from the media [19]. The changes progressed with age and consisted of fraying of the internal elastic lamina, penetration of muscle cells into the intima and formation of elastic tissue between the original lamina and the endothelium. Further development led to intimal thickening which ran longitudinally between the orifices of the arterial branches. The intimal thickenings were found to be more common and widespread with age and resembled the early discernible lesions in humans.

A number of different lesions forming a range of morphological variations were observed in these studies. In the susceptible swine there was an increase in number and severity with age. However, it is of interest to note contrasting differences with respect to the frequency of susceptibility to atherosclerosis. In one report all swine over 1 year of age, representing nine breeds, developed lesions which led to an inference that pigs as a species are inherently susceptible to the disease [25]. Other reports [59, 76], as well as subsequent studies on experimentally induced atherosclerosis, have shown that not all animals developed grossly identifiable changes [88, 97]. The incidence and degree of atherosclerosis varied from breed to breed [104] and at 8 years of age 50% of the swine, on the average, showed atherosclerotic involvement [97]. The most advanced lesions were observed in the oldest swine, which also showed the greatest variations in the susceptibility [59, 88].

Spontaneous atherosclerosis was further evaluated in pigs of the control groups of other experiments, including several strains of miniature swine. Their age varied, with the majority being close to 1 year of age. Approximately 40% of these pigs had few or none of grossly or microscopically detectable lesions [8, 30, 89]. The remaining 60% showed varying degrees of intimal thickening composed of connective tissue similar in general structure to that described earlier [29]. Different degrees of changes were observed, and the most advanced lesions appeared to have a familial (genetic) basis [96].

In summary, research data on spontaneous atherosclerosis in swine indicated that extensive advanced changes with complicated atherosclerotic lesions such as ulceration, thrombosis and hemorrhages into the lesions, identified in humans, have not been observed in swine. However, gross and microscopic examinations of the arteries in swine revealed many characteristics which appeared to be comparable to those in humans. This inference was based on the following indications: (1) a number of examined swine showed inherent susceptibility to atherosclerosis; (2) the same arteries (aorta, coronary and cerebral) were severely affected by atherosclerosis in swine as in humans; (3) swine atherosclerosis began early in life as fatty streaks or intimal thickenings; (4) regions of intimal thickening were common in the coronary arteries, and changes identified as early atherosclerosis occurred in these regions; (5) the changes showed similar distribution at points of branching as the lesions in humans and they increased progressively with proliferation of elastic tissues with age; (6) the lesions occurred when the animals consumed low-fat diets; (7) the changes were located first in the thoracic segment of the aorta, then in the abdominal portions, and developed later in the coronary arteries; (8) smooth muscle cells appeared to proliferate and migrate into the intima where plaques developed; (9) plaques had collagen, elastic tissue, smooth muscle cells, fibroblast-like cells, fat-filled cells, and exhibited fibrous tissue proliferation. Larger lesions with considerable amounts of extracellular fat had the appearance of atheromatous softening.

Dietary Induced Atherosclerosis in Swine

The observations that spontaneous arterial lesions in swine resemble the human disease stimulated interest to investigate the effects of different dietary ingredients, especially lipids alleged as atherogenic, their levels of intake and interactions on the development and acceleration of experimental atherosclerosis. Additional variables which could be associated, or indicative of the disease process, were considered in some of these studies and included standard breeds [2–4, 14, 30, 60, 95, 96] versus miniature swine [8, 35, 36, 47, 70, 89, 91, 103], age [2, 3, 35, 47, 89, 95, 96], sex [14, 30, 95, 96] and castration [47] as well as the effect of diets on changes in serum lipids (mainly cholesterol) and lipoproteins, and changes in arterial morphology and composition [17, 53]. The primary objectives were to search for indications that would elucidate the relationships between the type and intake of the diet on the development and process of the disease.

Although dietary fats have always been regarded as hypercholesterogenic and atherogenic factors, and therefore were intensely studied variables, the effect of dietary proteins and carbohydrates on experimental atherosclerosis has received some attention [2, 3, 30, 70, 103]. A single study, which used 24 young males of Pitman-Moore miniature swine, was carried out to test the influence of a long-term intake of moderate amounts of simple and complex carbohydrates on the concentration of serum lipids and atherosclerosis [103]. Feeding cholesterol-containing diets for 2 years, in which 60% of the calories were either from sucrose or cornstarch, showed no significant effects on serum cholesterol concentrations. The extent of atherosclerosis varied among the swine and appeared to be associated with the serum cholesterol concentration rather than the carbohydrate diets. Neither 1.1% content of cholesterol derived from egg yolk, nor the addition of 1% of crystalline cholesterol (total 2.1%) to the carbohydrate diets produced serum hypercholesterolemia except for the two responding animals.

The influence of dietary proteins and interactions between dietary proteins, fats and cholesterol on the serum cholesterol [2, 3, 30, 70] and atherosclerosis [30, 70] was the subject of studies in which older Yorkshire females [2], young boars [3], young crossbreds of both sexes [30] and young Pitman-Moore miniature swine of both sexes [70] were used. The results of these studies, with one exception [3], showed no significant effect of the protein content on the serum cholesterol level [2] and development of atherosclerosis [30,70]. Feeding 10–15% fats in the form of beef tallow [3,30], soybean oil or lard [30], markedly [3] or slightly [30,70] increased serum cholesterol. The addition of 1% cholesterol to the saturated fat diet resulted in a more than twofold serum cholesterol elevation and increased the incidence and severity of atherosclerosis [70].

The remaining study [2] which used a human-type, or a purified diet with varying amounts and types of fat and protein showed that saturated fat in the form of 7% butter increased serum cholesterol level more than unsaturated fats. This is in agreement with the earlier study [4] in which a control standard diet and two experimental isocaloric diets, one containing 22% corn oil and the other 26% butter, were used. Only the butter diet led to the rise of serum cholesterol; however, this elevation was not associated with an increase of atherosclerotic changes. Since the frequency and severity of lesions were the same in all three dietary groups, the susceptibility seems to have a genetic instead of dietary basis. Similarly a lack of advanced atherosclerosis was observed in another study [8] designed to test whether or not a typical American diet, rich in fat (a military ration), differs in atherogenicity from two synthetic diets, one described as a 'poor' (atherogenic) and the other as 'good' (balanced low-fat diet). Due to a low response to the diets and similar frequencies of lesions within the dietary groups, there was no correlation between the diets and atherosclerosis or cholesterol levels.

Contrasting to these reports [2–4, 30, 70, 95] was the result from a study [89] designed to determine the response in the level of cholesterol and other lipids in blood plasma, liver, aorta and coronary arteries to three diets; a low-fat, a high saturated and a high unsaturated fat diet. Only the high-fat diets with 2% added cholesterol induced atherosclerotic lesions; however, the high unsaturated fat with the cholesterol produced the most marked atherosclerosis. The high-fat diets with or without 2% cholesterol elevated tissue cholesterol, and the high unsaturated fat increased cholesterol absorption. Consequently, the high unsaturated fat without cholesterol produced higher cholesterol levels in liver and plasma, but not in the arteries, than did the high saturated fat. It was concluded that the plasma cholesterol level is not correlated with its level in the coronary arteries.

Consistent with this conclusion, however of the contrasting nature (i.e., dietary fat affecting atherosclerotic lesions but not serum cholesterol levels), are observations from two of the three studies [14, 95, 96] by the same group of investigators who compare the effects of fat-enriched diet on the development of atherosclerosis [95, 96]. In the first study [95], 40% of the calories of a standard low-fat diet (control) were replaced by either margarine or butter. The margarine diet produced very little increase in atherosclerosis, whereas the butter diet accelerated the progression of lesions. However, neither diet increased serum cholesterol or phospholipids. In the second study [96], 33% of the calories of the control diet were replaced by either butter or egg yolk, the latter assumed to be atherogenic in humans. The swine fed the egg yolk diet exhibited elevated serum cholesterol and six times more aortic atherosclerosis than controls and 50% of the animals showed coronary artery involvement. The butter diet increased three times the lesion involvement; however, similarly with the first study [95] it had no effect on the cholesterol level. However, contrasting effects in the response to feeding butter were observed in other studies [2, 4].

In view of the lack of effect of the butter diets [95, 96] on the serum cholesterol level, the third study [14] was carried out to investigate effects of dietary cholesterol and fat contents on the serum lipids and atherosclerosis. In each of the three fat-enriched diets, 25% of the calories were replaced by either egg yolk, lard, or lard plus cholesterol, the latter in the amount equal to that in the egg yolk diet. The comparison showed that lard alone was only slightly more atherogenic; egg yolk twice more, and lard and cholesterol diet two and half times more atherogenic than that found in the control group. Similarly as shown in other studies not all swine used in these three investigations developed lesions and the extent of changes among the susceptible animals exhibited marked variations. Several littermates were equally susceptible regardless of the diet even when consuming the standard swine ration low in fat. This variation in susceptibility was interpreted as a genetic trait [96].

In agreement with these observations [14, 96] on the effect of egg yolk on the elevation of serum cholesterol and development of arterial lesions are results from a study which also used young pigs [47]. The latter investigation [47] was designed to determine the effect of castration, or sham operation of male Hormel miniature swine fed a control or an atherogenic diet over a period of 1 year. Castration alone was accompanied by significantly elevated serum cholesterol. The castration and feeding the egg yolk-lard diet seemed to act synergistically resulting in hypercholesterolemia and increased fatty streaking. No correlation was found between the concentrations of plasma testosterone and serum cholesterol.

Contrasting with these observations [14, 96] are results obtained when the effects of egg yolk feeding on preexisting lesions were investigated [60] in five aged sows (12–14 years) derived from the herd used to study spontaneous atherosclerosis [59, 88]. An addition of 450 g (35% of caloric intake) of dry egg yolk per day to the customary diet [59, 88] for 10–14 months failed to accelerate aortic, coronary and cerebral atherosclerosis. Rather, the reversal of the naturally occurring atherosclerosis was observed in the aorta and the coronary arteries when the results were compared with those from five control swine fed only the garbage diet [59].

Another set of lipid variables, tallow and coconut oil [35] or tallow and cholesterol [36] were used to investigate the development of atherosclerosis [35], or severe atherosclerosis [36]. In the first study [35], swine fed 25% tallow with 1% cholesterol produced moderate response and slightly more changes than the coconut oil diet equivalent in the calories. In addition, marked variation in the susceptibility to the development of dietary induced lesions occurred among these swine. Discussion of the second study [36] will follow.

Summarizing the brief review of selected studies on the dietary induced atherosclerosis in swine, it appears that manipulation of ingredients in the diet had effects in the majority of experiments on the development of atherosclerosis. Diets enriched in saturated fats had accelerating effects, except for two studies [4,8], whereas, carbohydrates [103] and proteins [30, 70] showed no influence on the development of lesions. The extent of the disease was influenced by the types and contents of dietary lipids [30, 35, 70, 89, 95, 96]. The addition of cholesterol to the diet accelerated further the atherosclerotic process up to eightfold [97], increasing its rate and severity [14, 30, 35, 47, 70, 89, 96]. Equally important was the demonstration that the dietary-induced atherosclerosis in swine closely resembled the naturally occurring form in humans with regard to patterns, morphology and topography, however, with one essential difference. Unlike the spontaneous form [19], the experimentally induced disease was associated with some [35, 96] or marked serum cholesterol [30, 36, 47, 70, 89, 96] and lipoprotein elevations [24, 67, 90, 91, 96, 103]. This difference could be interpreted as follows: the elevated cholesterol is an unlikely initiator of the disease but its elevation is associated with the acceleration of atherosclerosis [70]. This relationship does not appear to be of a simple nature, since the diets with unsaturated lipids did [8, 89] or did not [95] influence serum cholesterol levels having no effect on arterial changes [8] with one exception [89].

While sex showed no effect on atherosclerosis [14, 30, 95, 96], castration increased the susceptibility [47]. The response to dietary manipulation seemed to undergo considerable changes with age. The young pigs appeared susceptible [35, 89, 95, 96], the older less responsive [4], and the aged swine quite resistant to the effect of atherogenic diets [60]. The matter of the standard size versus miniature swine was considered and seemed to favour miniature swine which were proved suitable and very economical breeds for studies of atherosclerosis [17, 70]. However, the Yorkshire breed on this continent [95, 96] and Large White swine as its counterpart in Europe [104] were considered as the most susceptible [92, 104], and the Landrace as more resistant than the miniature swine to the development of atherosclerosis [35, 39, 104]. Finally, results of these studies indicate that swine, like other species, showed a varying degree of susceptibility to the effect of atherogenic diets. However, the possible nature of these variations has not been studied or addressed except for a single comment [96].

Accelerated Atherosclerosis and Hypercholesterolemia

The main information supporting earlier indications that the accelerated atherosclerosis in swine is associated with plasma hypercholesterolemia came from studies undertaken following the lead that the experimentally enriched diets in lipids, by addition of cholesterol and high levels of human dietary fats, accelerate atherosclerosis. Some of these studies were continued to test the responses of different strains of swine [21, 23, 28, 31, 36, 91] to lipid diets [28, 36, 72] containing varying amounts of cholesterol (from 1.2 to 4.2%) [17, 53] and high levels of the human dietary fats (up to 66% of the calories) [54, 55, 107], or to diets combined with either X-irradiation [54] or with intimal injury by balloon catheterization [31, 107], in an attempt to produce advanced atherosclerosis in a shorter period of time in young pigs [36, 54]. Other studies in which dietary fat and cholesterol were used to obtain severe or advanced atherosclerosis were concerned with: (1) plasma lipoprotein changes [28, 37, 62, 67, 72, 90, 105]; (2) roles of monocyte-macrophages, lesion-prone areas of the aortic arch [22–24], and intimal cell masses (ICM) of the abdominal aorta [48, 53, 99, 100, 107] in the development of early atherosclerotic lesions, and (3) susceptibility and resistance to coronary atherosclerosis of swine with von Willebrand factor [21, 31]. Findings of these studies suggested, as it is universally recognized today [62], that severe dietary induced serum hypercholesterolemia promotes the development of advanced atherosclerosis [31, 36, 53, 107].

As shown in earlier investigations, there was a high degree of variability in the atherosclerotic lesions and the plasma cholesterol levels among swine, regarding the magnitude of response to the dietary fat and cholesterol. Such variations were of common occurrence in other species, including man. These variations extended from nonresponders to severe responders and some swine did not respond even to 3 or 4.2% cholesterol [17, 62].

Feeding hyperlipidemic diets led to the following changes in the responder swine: an increase in serum cholesterol observed after 2 weeks of feeding [24, 48], which progressed usually within 8 weeks to moderate (sixfold) [24, 28, 62, 67, 72, 91, 105], or to severe (elevenfold) hypercholesterolemia [31, 36, 48, 99, 107] in the hyperresponder swine. The serum cholesterol increase preceded detection of early proliferative lesions observed after 30 [17] or 49 days [48], and foam cell lesions [24] after 12 weeks. The proliferative lesions increased to high frequencies after 160 days [17]. The advanced lesions observed between 160 and 330 days of feeding were characterized by marked intimal proliferation, cholesterol crystal deposition and necrosis [36, 62, 91]. More advanced lesions appeared as extensive foci with one-third of the mass as lipid-rich calcified necrotic debris, which resemble those in humans consuming high-fat-cholesterol diets in this country [107]. The combined ballooned and high-lipid diets did not increase either cholesterol or severity of lesions [31, 107]. The severe atherogenic diet combined with X-irradiation elicited the most advanced atherosclerosis accompanied by many myocardial infarcts and sudden death [54].

Accelerated Atherosclerosis and Some Morphologic and Metabolic Factors in the Arterial Wall and Plasma

The development of models for accelerated atherosclerosis in swine provided opportunities to study mechanisms of atherogenesis and accompanying events in a short time span, which may occur during the entire period of life in man. Morphologic and metabolic factors and mechanisms recognized to play a part in the process of development of atherosclerotic lesions have been studied extensively [17, 107]. Studies concerned with early pathogenesis of atherosclerosis in the lesion-prone areas of the abdominal aorta and coronary arteries of young pigs, fed high-fat-cholesterol diets, as well as changes in these areas with aging in swine (3–12 years old) fed nonatherogenic diets, were performed by scientists at the Albany Medical College. The published data are too numerous for review here. The reader is referred to several review articles summarizing these studies [53]. More recently this group investigated aspects of endothelial cell alterations and the role of plasma monocytes during the initiation and progression of lesions in the disease-predisposed area of young pigs fed hyperlipidemic diets for 0, 14, 49 and 90 days [48, 99, 100].

Results of these studies are in agreement with human data and indicate that atherogenic lesions in both young and old swine occur in the lesion-prone area of the distal portion of the abdominal aorta and proximal part of the coronary arteries where the intima is thickened, from birth onwards, by multiple layers resulting from accumulation of cells, of which 90% are smooth muscle cells (SMC) and 10% monocyte-like cells, and extracellular elements such as collagen, elastic tissue and glycosaminoglycans. This thickened intima is referred to as intimal cell masses (ICM). By 90 days of feeding the atherogenic diet there was an eightfold increase in ICM, SMC and monocyte cells. This corresponded with: (1) lipid accumulation in SMC, ICM and monocyte-macrophages; (2) an increase in size of ICM lesions and appearance of a few foci containing lipid-rich calcific necrotic debris, and (3) an increase in leukocyte adherence to the endothelial surface associated with increasing turnover of the endothelial cells. Less advanced but similar changes were observed in the aged swine fed the standard low-fat diet. This similarity supports previous observations suggesting that the susceptibility to atherosclerosis is independent of feeding the atherogenic diet; however, the diet greatly accelerates the disease process.

Another lesion-predisposed area of swine aorta and mechanisms of atherosclerotic lesion formation were identified by Gerrity and coworkers [22–24, 37, 53]. Using young swine fed an atherogenic diet, it has been shown that areas of the aorta which had enhanced permeability to albumin and fibrinogen, and was demarcated by uptake of the protein-binding azo dye, Evans blue, accumulated LDL and was predisposed to early atherosclerosis [24, 37]. These lesion-prone aortic areas showed preferential adherence to the endothelium by blood monocytes which subsequently migrate into the intima and become the major source of foam cells in the early lesions [22]. The foam cells with accumulated lipids migrate from the lesion into the lumen. This suggested that the monocytes, which phagocytize and accumulate lipids, play a role in lipid clearance in early atherosclerosis. Prolonged hypercholesterolemia may overload the lipid clearance system, triggering SMC involvement, lipid accumulation, cell necrosis and subsequent lesion progression by a number of associated events and factors [53]. Studies on mechanisms of monocyte migration into the lesion-prone areas revealed that extracts of blue areas from the hypercholesterolemic pigs, but not from normal-diet fed animals, had factors chemotactic only for monocytes from the hypercholesterolemic pigs.

Swine with von Willebrand's disease (vWD) have inherited disorders affecting platelet function which mediate platelet-vessel wall interactions. This defect makes vWD swine a useful model to test the hypothesis that impaired platelet response to vascular injury leads to a reduction in the incidence of high cholesterol diet-induced coronary atherosclerosis. Two studies were reported recently on this subject and their results are inconsistent [21, 31]. In the first study of coronary atherosclerosis in normal and vWD swine, very little coronary disease was detected in either spontaneous or high cholesterol diet-induced atherosclerosis [21]. Only one pig of each phenotype in the diet-induced study developed moderate luminal reduction. In the second study [31] the development of coronary atherosclerosis in response to acute intimal injury, by balloon catheterization, and dietary induced hypercholesterolemia in normal and vWD swine, no difference in the coronary atherosclerosis were observed between ballooned and nonballooned vessels within either phenotype. The major variable affecting coronary atherosclerosis was serum cholesterol which showed significant positive correlation with the intimal lesions ($r=0.62$; $p=0.006$). The absence of a synergistic effect of balloon injury and hypercholesterolemia on the development of proliferative atherosclerosis in this study differs from earlier observations. It was concluded that a proliferative effect of lipid on coronary lesion development is independent of a major influence of platelet adhesion and release [31].

Hereditary Aspects of Hyperlipidemia and Response to Atherogenic Diets

Coronary heart disease is the leading cause of human deaths in the United States and a number of factors correlated or associated with the risk of coronary heart disease have been identified. Unfortunately, a large proportion of the incidence of coronary heart disease is not explained by the classic risk factors. Among the classic factors associated with the risk for atherosclerosis is hypercholesterolemia which is highly correlated with elevations of LDL concentrations. Epidemological and population genetic studies on the causes of cholesterol and LDL elevations revealed that dietary induced hypercholesterolemia and atherosclerosis are common in humans as well as in research animals, and have been presumed to be genetic [13].

As indicated in this review, variations in the susceptibility to the development of atherosclerosis and hypercholesterolemia in response to the atherogenic diet were also common in swine; however, the nature of these variations has not been addressed. A number of other studies, including those on swine [58, 71, 94], revealed that the genetic factors exert effect on the cholesterol and LDL concentrations [5, 18, 32, 38, 57, 101, 105, 108]. Except for relatively rare cases of monogenic hypercholesterolemia in humans [5] and rabbits [108], resulting in both species from defects in LDL receptor activity, the mechanisms as well as the molecular basis remain unexplained. Genetically mediated LDL and cholesterol elevations may result from the action of a number of genes, diet-genotype interactions, or interactions between products of two or more genes or genotypes, which carry similar or related physiological functions in metabolic processes of lipoproteins or lipids. Polygenic abnormalities in lipid and LDL metabolism may be common, but are difficult to identify without gene-specific markers. This is further complicated by more recent indications that all plasma lipoproteins are metabolically related and are exposed to different lipid enzymes and lipid transfer proteins [101].

Biochemical studies during the last decade have shown that apolipoproteins and enzymes, active at lipid interfaces, play a central role in lipid transport systems and lipid metabolism. Advances in our knowledge on lipid transport and metabolism have been considerable. Recent studies on various aspects of plasma lipoproteins, including physicochemical characterization, structure, metabolism, cell biology and molecular biology, have contributed considerably to a better understanding of lipoproteins [for reviews, see 64, 101]. Prior to that, little progress was made in biochemical analysis of polypeptide composition and identification of mutations in the main plasma lipoprotein, LDL, due to poor solubility of its principal protein, apolipoprotein B (apo-B). Another contributing factor was insufficient interest in and support for exploring genetic aspects of lipoproteins. A member of the 'Task Force on Genetic Factors in Atherosclerotic Disease' [106, p.179] stated that the nature of lipoproteins 'is not currently being approached from the aspects of genetics'.

Plasma Lipoproteins in Swine Fed Standard Low- or High-Fat and Cholesterol Diets

Lipoproteins of all the species studied are water soluble, heterogeneous, and the most complex blood plasma macromolecules, composed of lipids (cholesterol, phospholipids and triglycerides) and specific proteins termed apolipoproteins. They are classified on the basis of their properties such as density, size, electrophoretic mobility, apolipoprotein content, gel chromatography or affinity chromatography separations. From published data on a number of lipoprotein characteristics, including concentration, distribution, composition, metabolism, structure, heterogeneity, dyslipoproteinemia and physiological function, it is evident that lipoproteins represent an important biological system in lipid transport and metabolism. Through these attributes, lipoproteins are strongly implicated in atherogenesis. The essential part of the lipoprotein is the apolipoprotein which maintains the lipoprotein structure, regulates its metabolism, transports and redistributes lipids among various tissues and acts as cofactor for the main lipid enzymes [for review, see 65]. There are at least 12 apolipoproteins identified in humans [1].

Swine plasma lipoproteins have been studied in normal physiological conditions [10, 41–44, 46, 49, 66] and under experimentally induced changes of the metabolism by hypercholesterolemia, induced by feeding high-fat and cholesterol diets [9, 36, 62, 67, 72, 90, 91, 105]. Due to a large number of studies reporting on various characteristics of swine lipoproteins, only selected aspects, such as resemblance to human lipoproteins and dietary induced changes, will be reviewed here. Some other aspects which were the subject of our studies will follow. The reader is referred to two more recent reviews, one on swine lipoproteins exclusively [40] and the other on swine and other animals [11].

Early investigations on the comparison of several mammalian lipoproteins revealed that swine bear the closest resemblance to human with regard to distribution and composition of plasma lipoproteins [34]. Later studies showed that swine serum contains similarly to humans the three major classes: very low density (VLDL), low density (LDL) and high-density (HDL) lipoproteins, which can be separated by ultracentrifugation, electrophoresis or gel filtration [43, 44, 46, 66]. Plasma cholesterol levels in fasted pigs, maintained on standard swine diets, showed considerable variations of 60–73 [28, 90, 99], 89–90 [66, 67, 72], 105 [34], 112 [24] and 150–230 [49] mg/dl, but were found at considerably lower levels than in humans. The most extensive comparison of human lipoproteins with swine by two methods (electrophoresis and standard ultracentrifugation) showed that both lipoproteins are essentially identical with respect to chemical composition, immunochemical reactivity, size by electron microscopy and apoprotein content by polyacrylamide gel electrophoresis [66]. The observed differences between swine and human lipoprotein were: (1) extended density distribution in swine to d 1.09 g/ml [43, 44]; (2) an absence or very low concentration of apo-AII [41], and (3) prominence of apo-CII in VLDL [49]. The presence of two LDLs in swine, LDL1 and LDL2, was confirmed by several studies [10, 42, 49, 66].

Feeding atherogenic diets high in fat and cholesterol, which led to moderate [24, 28, 62, 67, 72, 91, 105] or severe hypercholesterolemia [31, 36, 48, 99, 107] and accelerated atherosclerosis, was always correlated with LDL elevation [24, 67, 90, 91, 96, 103] and caused dramatic changes in the distribution and size of LDL1 and LDL2 [72]. Extensive studies on miniature swine performed by Mahley and co-workers [62, 67, 72, 90, 91, 105], contributed to the finding that feeding high fat and cholesterol induces marked changes in plasma lipoproteins, which are similar to both qualitative and quantitative changes in the various species including humans [11, 63]. These changes include: (1) the appearance of B-VLDL in the $d < 1.006$ g/ml fraction and a lipoprotein referred to as HDL$_c$

(cholesterol-induced); (2) increases of LDL and intermediate density lipoprotein (IDL, d 1.006–1.02 g/ml), and (3) a decrease in typical HDL [62, 67, 90, 105]. The B-VLDL lipoprotein is a β-migrating, cholesterol-rich VLDL, and the HDL_c has α-mobility of d 1.02–1.087 g/ml, and, while lacking apo-B, is a cholesterol-rich lipoprotein containing apo-E, apo-Al and the fast migrating C apoproteins [67]. These two lipoproteins, β-VLD and HDL_c, transport primarily dietary induced plasma cholesterol and were termed atherogenic lipoproteins. Contrasting with the presence of HDL_c in the atherogenic diet-induced plasma hypercholesterolemia is an absence, or very low levels of HDL_c in the plasma of our recently developed strain of swine with inherited hyper-LDL-cholesterolemia (IHLC) [78] which will be discussed later. In the IHLC swine, elevated cholesterol is associated primarily with the buoyant fraction [12], or layers 2 and 3 of the gradient ultracentrifuge fractions [51,52].

Immunogenetically Identified Lipoprotein Polymorphism in Swine Serum

Newly developed immunological procedures by use of agar gel in the visual detection, by precipitation, of serological reactions between plasma antigens and corresponding antibodies facilitated, among other applications, many successful attempts in detecting distinct variants of a single protein. If the coding structural gene for the protein was large, a number of mutations were likely to produce a series of mutated forms of the gene (allelic genes), each coding for a modified corresponding protein structure, and this process gave rise to genetically evolved polymorphisms. The modified part of the protein, if antigenic, could elicit an antibody for this variant (allotype, epitope, antigenic specificity) in animals of the same species which lack this allotype.

Without aiming at any specific blood plasma protein, pooled whole swine sera were inoculated into 22 pigs two decades ago, assuming that the antigenically strongest and most polymorphic protein will elicit antibodies. One of the strongest reacting alloimmune serum produced precipitation patterns of four discernible phenotypes which were finally resolved into two closely genetically and immunochemically related allotypes, designated tentatively as Lpp1 and Lpp2. The allotypes were found to be associated with lipoproteins of hydrated density, d 1.002–1.075 g/ml, and determined by two codominant autosomal allelic genes, Lpp^1 and Lpp^2, or less likely, by two tightly linked genes [77].

This discovery, reports on identification of several LDL allotypes in human plasma (Ag) [7], and anticipation of the importance of LDL in lipid transport and metabolism stimulated further interest in exploration of swine lipoprotein polymorphisms. Using LDL fractions as alloantigens, immune sera were obtained detecting four new allotypes, which exhibited identical physicochemical properties as the original two Lpp allotypes and were designated as Lpp3, Lpp4, Lpp5 and Lpp15 [79, 82]. Immunochemical and genetic studies, and a survey on the distribution of the six Lpp allotypes in over 6,000 pigs, derived from 34 breeds, suggested that five allelic genes, Lpp^1, Lpp^2, Lpp^3, Lpp^4 and Lpp^5 control the following five phenogroups: Lpp1,15; Lpp2,15; Lpp3,15; Lpp4,15, and Lpp5. Common occurrence of Lpp15 in the four Lpp phenogroups led to name this epitope as a common marker in contrast to the other five, Lpp1–5, designated as individual or mutant allotypes. Using lipoprotein preparations of $d < 1.085$ g/ml, as antigens, derived from various breeds of swine, additional alloimmune sera have been obtained during the first 10 years of investigations, and at least 16 monospecific anti-Lpp alloreagents were obtained, each producing different reaction patterns with normal plasma of a distinct Lpp genotype as shown in figure 1a.

Immunochemical characterization indicating that the Lpp allotypes are associated with the main LDL protein, apo-B, was confirmed by immunoblotting experiments [84]. The original nomenclature for the LDL allotypes was changed from Lpp to Lpb (L, lipoprotein; p, pig; b, apo-B or $apo-B$ locus) [75]. Thus these experiments confirmed our previous contention that there is an extensive epitope polymorphism associated with the most complex apolipoprotein, apo-B.

Extended genetic studies, carried out on over 10,000 swine, representing over 30 breeds, have shown that the Lpb allotypes are inherited in groups of eight epitopes [74, 75, 78]. Data from the double immunodiffusion test, devised for studying the phenotypic expression of Lpb alleles, have indicated that the sets of Lpb epitopes are inherited as genetic units, occurring together on distinct LDL particles [80, 85]. These results served as the basis for the construction of the Lpb model [74] and led to a conclusion that the set of eight epitopes, as shown in figure 1, is encoded by a single codominant allelic $apo-B$ gene [74, 75]. A total of eight sets (phenogroups, haplotypes) have been identified, as shown in the model. Each set has seven common epitopes, designated Lpb11–Lpb18, and differ from each other by the eight distinctive characterizing epitope, named individual or mutant epitope, designated Lpb1–Lpb8, except for the last set, which has two mutant allotypes, Lp5 and Lpb8.

No exception has been encountered with regard to the number and epitope specificity in the Lpb phenogroups among almost 15,000 swine surveyed. Each of the 16 anti-Lpb reagents has its unique value in studying the genetic complexity of apo-B, as well as in identifying new Lpb

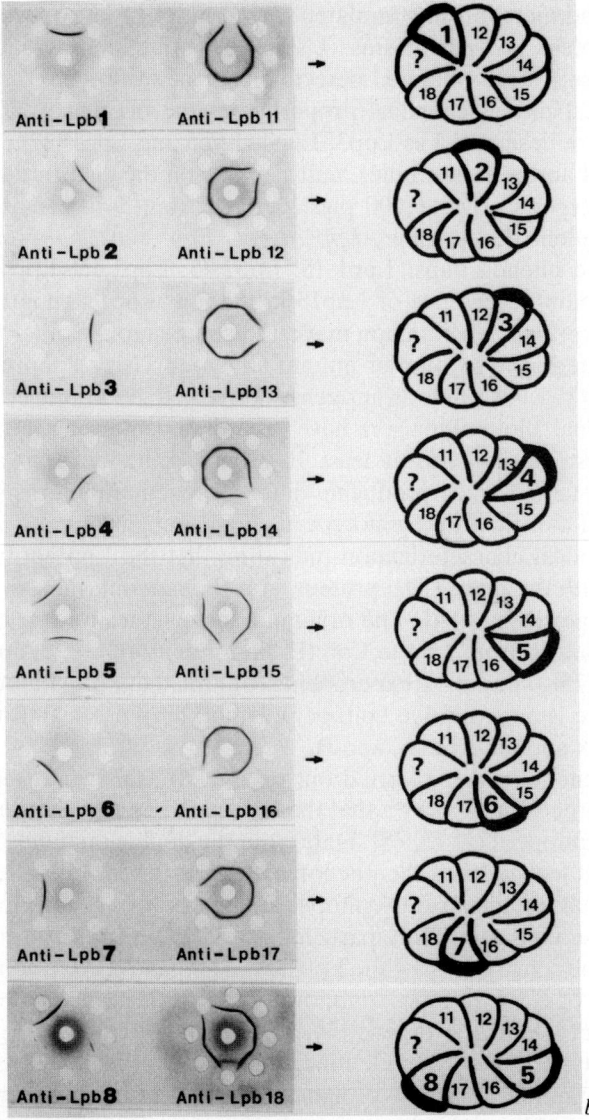

Fig. 1. Transformation of patterns of serological reaction into a schematic model of the apo-B-bearing lipoproteins, designated Lpb.

a Patterns of precipitation reaction in agar gel of eight plasma sera (peripheral wells), representing eight Lpb homozygous genotypes, *Lpb¹ᐟ¹-Lpb⁸ᐟ⁸* placed clockwise starting at the top of each rosette, respectively, reacted with 16 alloimmune reagents, anti-Lpb1-Lpb8 and anti-Lpb11-Lpb18, as indicated. Each reagent detects a different epitope in the apo-B structure. Each of eight epitopes, designated as individual (mutant) allotypes, Lpb1-Lpb8, mutants [74, 80]. Except for two very common alleles, *Lpb⁵* and *Lpb⁸*, others are found at low frequencies and their distribution is associated with specific breeds [73]. Since no exception was found with regard to the number and specificity of epitopes in each set, we proposed that the original *apo-B* gene in swine contained genetic information only for the common allotypes, Lpb11–Lpb18. The individual epitopes appeared later during speciation, when single mutations at various parts of the *apo-B* gene took place, leading to the appearance of the code for the individual allotype, and the loss of information for the corresponding common epitope. This process resulted in the formation of mutually exclusive pairs of epitopes (e.g., Lpb1 is the substitute for Lpb11, or Lpb7 for Lpb17 in the respective haplotype, and there are eight pairs of mutually exclusive allotypes identified). Although no comparable apo-B epitope complexity to swine was found in other species, rhesus monkeys and chicken closely resemble the swine model [33,75]. The genetic control of 10 human LDL markers designated Ag allotypes was proposed to be by five pairs of allelic genes placed at closely linked loci [7], although a single locus with a multiple series of alleles cannot be excluded [for review, see 74].

In addition to the 16 Lpb epitopes, six other lipoprotein allotypes have been reported by this laboratory, and include: Lps1, Lpt1, Lpr1, Lpr2, Lpu1 and Lpu2. Each allotype is determined by the individual gene; two autosomal dominant *Lps¹* and *Lpt¹*, and two pairs of autosomal codominant alleles, *Lpr¹*, *Lpr²* and *Lpu¹*, *Lpu²*, respectively. Homology of Lps1, Lpt1, Lpu1 and Lpu2 allotypes with any of the human apolipoproteins [1] is unknown, however the Lpr lipoprotein is immunologically related to human apo-D [69]. Characterization of Lpr revealed that the apolipoprotein forms an allelic multimer, which is distributed mainly in very high density lipoproteins (VHDL). The remaining Lpr occurs in VLDL in complexes with β-VLDL [81]. Plasma concentrations of Lpr1 in

---

forms a mutually exclusive pair of epitopes with one of eight (Lpb11-18) corresponding common (ancestral) allotypes (e.g., Lpb2-Lpb12, or Lpb5-Lpb15). The Lpb phenogroup (haplotype) of each plasma is defined by the sum of the precipitation reactions with the Lpb reagents, e.g., the plasma in the top wells according to its reactions with the 16 reagents exhibits the Lpb1, 12, 13, 14, 15, 16, 17, 18 phenogroup. The gel slides with the developed reactions were processed and stained with Sudan Black B.

b Immunogenetic model of the Lpb lipoproteins in swine. Schematic representation of eight identified Lpb haplotypes (phenogroups) each corresponding to single Lpb (*apo-B*) allelic gene designated by numerals, 1–8. Each allele of apo-B determines a complex haplotype bearing a set of at least eight distinct apo-B epitopes.

the $Lpr^1/1$ pigs is three times higher than Lpr2 in $Lpr^2/2$ animals. The Lpr^1 allele was found so far at very low frequency and only in Chester White pigs.

All swine allotypes identified [75] are associated with the LDL class, except Lpr1 and Lpr2 [81]. The genetic loci for Lpt and Lpu are very closely linked with the apo-B locus. The Lpu^1 gene was found in a very small number of pigs, also of Chester White origin, and is linked with Lpb^5, forming the Lpb^5 Lpu^1 haplotype. The Lpt1 allotype may be associated with lipoprotein(a) [16], may represent a second apo-B like apolipoprotein [74], or an apolipoprotein not yet identified. The Lpt^1 gene was found linked with all Lpb alleles, except Lpb^3.

Lpb Polymorphism and the Response to Feeding High-Fat or Atherogenic Diet

Indications implicating Lpb alleles as genetic factors associated with either enhancing susceptibility or resistance to diet-induced choles- terolemia and/or atherosclerosis were obtained from two studies [74, 76]. Twelve of 22 Thompson miniature swine fed an atherogenic diet (25% fat, 2% cholesterol) for 7 months showed resistance to plasma cholesterol elevation while the remaining 10 swine were responders. The subsequent Lpb test revealed that the all resistant pigs were homozygous $Lpb^3/3$, whereas the responders were genotyped as heterozygous $Lpb^3/5$ [74]. In the second study [76] the effects of consuming a diet for 10 months, containing 19% fat with varying mixtures of $trans$-unsaturated, cis- unsaturated and saturated fatty acids were studied to determine whether the consumption of $trans$-unsaturated fatty acids by swine would induce more aortic lipidosis than saturated fatty acids. Aortas and lipids in plas- ma and various tissues were evaluated in 57 pigs which were crossbreds carrying components of four breeds. Statistical analysis of the data shows an effect of diet-Lpb genotype interaction ($p < 0.003$) on fatty streaking: pigs carrying the Lpb^5 allele had a greater tendency to develop fatty streaking than those with Lpb^8. The mechanisms underlying the associa- tions in both studies are unknown.

Lipoprotein Polymorphism and Inherited Hyper-LDL- Cholesterolemia (IHLC) and Accelerated Atherosclerosis

The first marked elevation in plasma concentrations of LDL was observed in 1975 when results of the single radial immunodiffusion

(SRID) [68] from testing our experimental swine herd were compared. Several plasma samples, derived from related Chester White swine and their crosses with Hampshires, had elevated LDL. The pigs were fed a cholesterol-free, low-fat (<4%) diet. Plasma cholesterol was also elevated in swine with the elevated LDL levels, and its values were highly correlated (r = 0.91) with the LDL estimates. Results of plasma LDL and cholesterol quantitation in offspring of the hyperlipidemic pigs also showed elevation; however, considerable variations were observed among them regarding LDL and total cholesterol, which varied from 10 to 120% above the average for the experimental herd (98.3 mg/dl). These observations led to a supposition that the basis for these elevations is genetic. To test this hypothesis, a group of 12 swine served as original breeders in our Immunogenetic Project Herd (IPH) to produce five generations with the total of 175 closely related pigs [78]. The cholesterol level increased through the generations reaching a mean of 266 mg/dl in the fifth generation. The mean cholesterol value for all five generations was 176.5 ± (SD) 63.5 mg/dl. During 6 years of these studies we observed that the animals which had the highest LDL and cholesterol levels were invariably homozygous $Lpb^5/5$ $Lpu^1/1$, and in addition they were heterozygous or homozygous for Lpr^1. Three additional generations were produced subsequently and the mean cholesterol value for a total of eight generations of $Lpb^5/5$, $Lpu^1/1$, $Lpr^1/1$ animals is 257 ± (SD) 67 mg/dl.

The final conclusion from these studies was that the genetically mediated spontaneous plasma low density hyperlipoproteinemia and hypercholesterolemia have been identified in a strain of swine developed in the experimental herd of the Immunogenetic Laboratory. The hyperlipoproteinemia phenotype is highly correlated with hypercholesterolemia and is associated with three lipoprotein allotypes, Lpb5, Lpu1 and Lpr1 [78]. The cause of this elevation, or mechanisms underlying the associations are unknown. Segregation data from the progeny test indicated that the inherited hyperlipoproteinemia and hypercholesterolemia (IHLC) have a polygenic basis. Whether the Lpu1 and Lpr1 mutant epitopes contribute to the hypercholesterolemia in the Lpb^5 pigs remains to be determined. The mean plasma cholesterol value for pigs of very common $Lpb^5/5$, $Lpu^2/2$, $Lpr^2/2$ genotype, which were derived from six different breeds, was 105.4 ± (SD) 12.8 mg/dl. The mean cholesterol value for pigs carrying apo-B alleles other than Lpb^5 or Lpb^8 in the Lpb genotype, was 81.3 ± (SD) 11.6 [78, 84]. The difference between the means for these two Lpb genotypes is statistically significant. These results are interpreted as an indication that the inherent association exists between some homozygous Lpb genotypes, especially $Lpb^5/5$, and the distinct cholesterol and LDL concentrations. This contention is further supported by

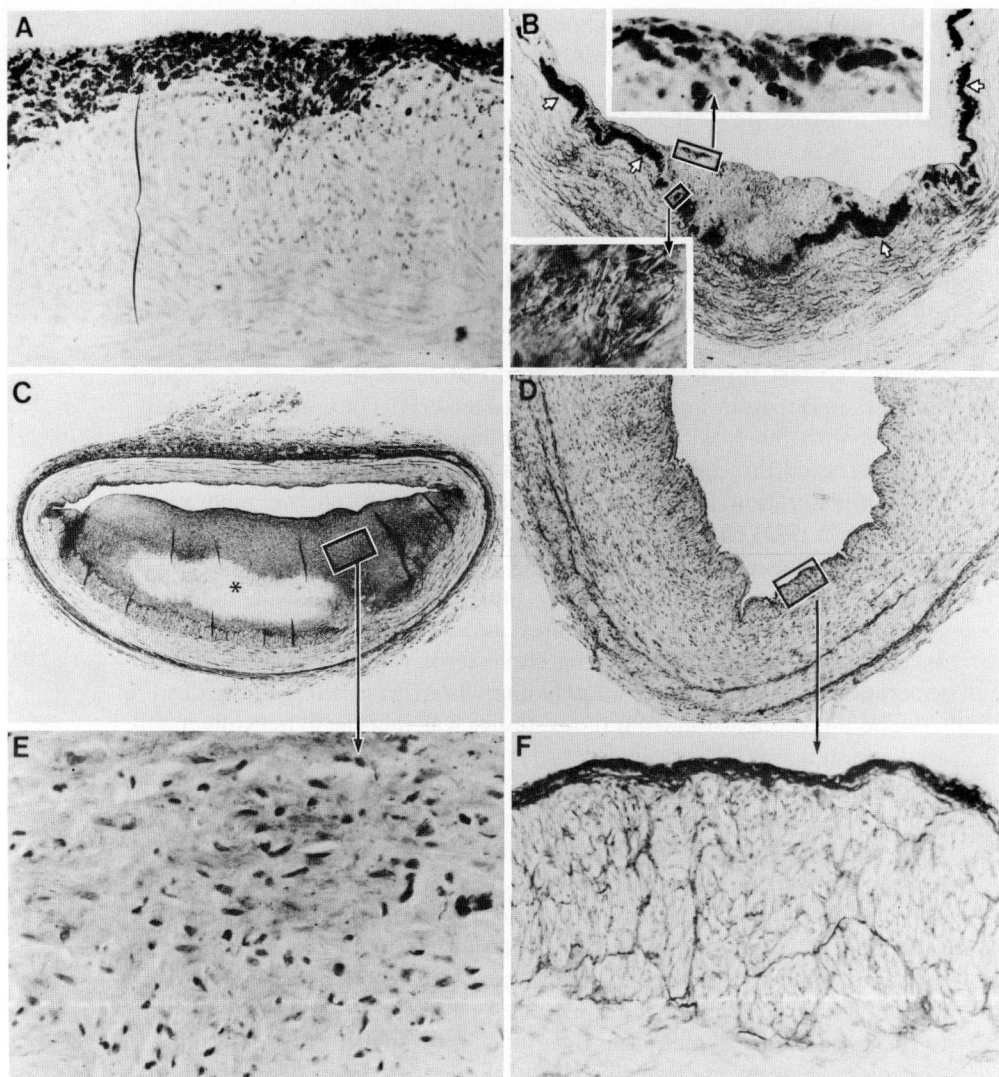

Fig. 2. Lesions in the right coronary artery of the $Lpb^{5/5}$, $Lpr^{1/1}$, $Lpu^{1/1}$ mutants compared with controls. *A* A typical fatty streak lesion from the right coronary artery of a 7-month-old mutant pig showing oil red O-positive neutral lipid contained within the cells. These lesions were composed of macrophage-like foam cells overlying an area of moderate intimal smooth muscle cell hyperplasia (bracket). *B* From a mutant pig at 21 months of age. Lipid is predominantly extracellular in the form of crystals 18–26 μm in length (lower inset), in the intima. Foam cells are a prominent feature of even the most advanced lesions (upper inset). *C* Distal portion of the same right coronary artery, stained with GTAF, shows a

a recent observation on cholesterol and LDL concentrations in another Lpb homozygote, $Lpb^7/7$ [83]. Due to a low gene frequency of Lpb^7, only 24 swine of the $Lpb^7/7$ genotype were evaluated; however, they all had the lowest cholesterol values, mean 69.8 ± (SD) 7.3 mg/dl [Rapacz and Hasler-Rapacz, unpubl.]. Although the cholesterol and LDL concentrations for the remaining Lpb homozygotes are not yet established, due to their low number, the observations on the relationships between the tested specific Lpb genotypes, LDL and cholesterol concentrations seem to indicate that the apo-B polymorphism reflects on lipoprotein and cholesterol metabolism in swine. This finding is unique and of biological importance for future studies on lipoproteins, lipids and cholesterol metabolism and atherosclerosis, as well as for elucidation of apolipoprotein mutations in relation to physiological functions of lipoproteins.

High rates of sudden death during the development of the IHLC strain of swine and a short life span (up to 4 years) prompted us to collect hearts for histological analysis to examine the arterial lesion involvement. Seventeen hearts from the hypercholesterolemic swine and seven from normolipemic animals, ranging in age from birth to 62 months, were collected and examined [84]. Pigs carrying the Lpb^5 mutant had extensive fatty streaks and foam cell lesions at 7 months of age (fig. 2A). Animals between 7 and 14 months of age developed extensive lesions in all three coronary arteries. At 21 months of age, complex and large lesions were common (fig. 2B) restricting blood flow in the affected swine (fig. 2C), while the right coronary artery from a 21-month-old control pig exhibited moderate smooth muscle cell hyperplasia in the media immediately below the elastic lamina (fig. 2D, F), a normal finding [19, 20, 29, 88, 97]. Unlike in familial hypercholesterolemia in humans [5] and rabbits [108] the IPH swine with the inherited hyper-LDL and cholesterolemia phenotype have normal LDL receptor activity, thus, IHLC highlights discovery of another genetic defect in lipoprotein and lipid metabolism. In all studies on dietary induced hypercholesterolemia and accelerated atherosclerosis in swine cited earlier, the hypercholesterolemia was of the exogenous

---

stenosing lesion with a necrotic core (asterisk) and a thin fibrous cap. *D* Right coronary artery in a control pig matched for age, sex, and heart weight showing the absence of lesions (stained with GTAF). *E* Cellular hyperplasia in the intimal compartment. This is abnormal, especially when it is associated with lipid infiltration, fibrohistiocytic cell populations, and loss of fascicular architecture. *F* Higher magnification of *D* showing moderate medial hyperplasia consisting of longitudinally oriented medial smooth muscle fascicles located immediately below the internal elastic lamina. Moderate smooth muscle cell hyperplasia in the inner media is considered normal in pigs [Reprinted from reference 84, with permission].

origin when the normal mechanisms of lipoprotein clearance were over-whelmed by very large amounts of dietary cholesterol, therefore making the dietary induced hypercholesterolemia model less desirable for studies of the human disease.

To better characterize atherosclerotic lesions in IHLC swine, im-munocytochemical, histologic and morphologic analyses were conducted on arteries of the two oldest IHLC females, 48 and 54 months of age. Complex lesions in the coronary arteries contained foci of the following: inflammatory cells, smooth muscle cells with extensive connective tissue deposition, macrophage-derived foam cells, necrosis with extracellular lipid and calcification, neovascularization and hemorrhage [93]. Although numerous studies showed close resemblance between swine lesions and human early lesions, new features observed in this study are indicative that advanced atherosclerosis in swine with IHLC resemble the human advanced disease. In our IHLC swine, accelerated atherosclerosis occurs when feeding low-fat, cholesterol-free diets and the elevated endogenous cholesterol is carried by catabolically defective lipoproteins. In conclu-sion, these IHLC swine represent a unique and valuable resource as an animal model for studying human atherogenesis.

Current Investigations on Swine with Inherited Hyper-LDL-Cholesterolemia (IHLC) and Accelerated Atherosclerosis

The development of a strain of swine with inherited LDL and cholesterol elevations [78], associated with accelerated atherosclerosis [84], stimulated further interest in characterization of LDL and its meta-bolism. Studies completed recently on chemical properties of LDLs [12], as well as on the profile and composition of lipoprotein subfractions from plasma of the IHLC swine showed [51, 52] that LDL exhibits a defective catabolism. Pigs with the Lpb^5 mutant LDLs have 30% lower catabolic rate in vivo and this accounts for hypercholesterolemia, abnormal com-position and distribution of LDL. This metabolic defect is associated with the larger, more buoyant cholesterol ester-rich lipoprotein particles than found in the LDL subfractions of normolipemic swine. It is conceivable that the accumulated larger more buoyant LDL in these pigs is especially atherogenic, leading spontaneously to the development of severe coro-nary atherosclerosis.

Because the IHLC swine showed normal LDL receptor activity, studies were undertaken to examine the activity of one of the major lipid enzymes, lecithin-cholesterol acyltransferase (LCAT), which regulates the plasma cholesterol levels catalyzing the formation of esterified cholesterol in the plasma of mammals [26]. Completed studies [50] on 42 IPH swine; 28 hypercholesterolemic and 14 normocholesterolemic showed that the LCAT activity was positively correlated with total cholesterol in the normocholesterolemic group ($r = +0.54$), whereas it was negatively correlated ($r = -0.82$) with total cholesterol in the hypercholesterolemic swine. Further study revealed that the lower LCAT activity in hypercholesterolemic pigs is at least in part due to LCAT inhibition by components of the lipoprotein-free fraction.

Finding in humans that elevated lipoprotein(a), Lp(a), is associated with coronary atherosclerosis stimulates interest in finding a suitable animal model. Very preliminary investigations using swine of different *Lpb* genotypes indicate the presence, variations in concentrations, and polymorphism of Lp(a) in swine. Plasminogen which is structurally related to Lp(a) is the inactive plasma protein precursor of the proteolytic enzyme plasmin, which plays a central role in the fibrinolytic system. Among identified polymorphic mutants in humans, a rare defective variant was found associated with recurrent thrombosis. Using anti-human plasminogen antibodies, different polymorphic forms were identified in 36 plasma samples derived from pigs of different Lpb genotypes [87]. Using anti-apo-B epitope-specific antibodies as immunosorbers, swine LDL from Lpb heterozygotes were separated recently into two apo-B haplotypes [85, 86]. These results confirmed previous studies on monogenic LDL molecular expression of Lpb in the Lpb heterozygous pigs. The existence of two heterozygous populations of molecules will make possible studies on metabolic differences at a single apo-B gene level of two mutant LDL populations.

Studies on the apo-B genes at the DNA level have been initiated by Maeda et al. [61] using two human cDNAs as probes for cloning experiments. Cloned portions of the pig apo-B gene were subsequently used to identify six RFLPs in the apo-B genes of pigs carrying different *Lpb* types. These studies have established that all eight immunogenetically identified [74] apo-B alleles are recognizably different at the DNA level.

Acknowledgments

The authors are grateful to Drs. A. B. Chapman and R. M. Shackelford for the editorial help and to Drs. J. J. Lalich and R. H. Grummer for their support and encouragement during the early period of these studies. We thank J. Busby for typing the manuscript. This research was supported in part by the College of Agricultural and Life Sciences and the Graduate School, University of Wisconsin-Madison, the Wisconsin Pork Producers and NIH AG 05856 and NIH HL 39774.

References

1 Alaupovic, P.: The role of apolipoproteins in lipid transport processes. Ricerca Clin. Lab. *12:* 1–21 (1982).

2 Barnes, R. H.; Kwong, E.; Fiala, G.; Rechcigl, M.; Lutz, R. N.; Loosli, J. K.: Dietary fat and protein and serum cholesterol. I. Adult swine. J. Nutr. *69:* 261–268 (1959).

3 Barnes, R. H.; Kwong, E.; Pond, W.; Lowry, R.; Loosli, J. K.: Dietary fat and protein and serum cholesterol. II. Young swine. J. Nutr. *69:* 269–273 (1959).

4 Bragdon, J. H.; Zeller, J. H.; Stevenson, J. W.: Swine and experimental atherosclerosis. Proc. Soc. exp. Biol. Med. *95:* 282–284 (1957).

5 Brown, M. S.; Goldstein, J. S.: A receptor-medicated pathway for cholesterol homeostatis. Science *232:* 34–37 (1986).

6 Bustad, L. K.; McClellan, R. O.: Swine in biomedical research (Frayn, Seattle 1966).

7 Bütler, R.; Morganti, G.; Vierucci, A.: Serology and genetics of the Ag system; in Peeters, Protides of the biological fluids, vol. 19, pp. 161–167 (Pergamon Press, Oxford 1972).

8 Calloway, D. H.; Potts, R. B.: Comparison of atherosclerosis in swine fed a human diet or purified diets. Circulation Res. *11:* 47–52 (1962).

9 Calvert, G. D.; Scott, P. J.: Serum lipoproteins in pigs on high-cholesterol-high-triglyceride diets. Atherosclerosis *19:* 485–492 (1974).

10 Calvert, G. D.; Scott, P. J.: Properties of two pig low density lipoproteins prepared by zonal ultracentrifugation. Atherosclerosis *22:* 583–599 (1975).

11 Chapman, M. J.: Animal lipoproteins: Chemistry, structure and comparative aspects. J. Lipid Res. *21:* 789–853 (1980).

12 Checovich, W. J.; Fitch, W. L.; Krauss, R. M.; Smith, M. P.; Rapacz, J.; Smith, C. L.; Attie A. D.: Defective catabolism and abnormal composition of low density lipoproteins from mutant pigs with hypercholesterolemia. Biochemistry *27:* 1934–1941 (1988).

13 Clarkson, T. B.; Prichard, R. W.; Bullock, B. C.; St. Clair, R. W.; Lehner, N. D. M.; Jones, D. C.; Wagner, W. D.; Rudel, L. L.: Pathogenesis of atherosclerosis: Some advances from using animal models. Exp. molec. Pathol. *24:* 264–286 (1976).

14 Downie, H. G.; Mustard, J. F.; Rowsell, H. C.: Swine atherosclerosis: The relationship of lipids and blood coagulation to its development. Ann. N. Y. Acad. Sci. *104:* 539–562 (1963).

15 Flaherty, J. T.; Ferrans, V. J.; Pierce, J. E.; Carew, T. E.; Fry, D. L.: Localizing factors in experimental atherosclerosis; in Likoff, Atherosclerosis and coronary heart disease, pp. 40–83 (Grune & Stratton, New York 1972).

16 Fless, G. M.; Rolih, C. A.; Scanu, A. M.: Heterogeneity of human plasma lipoprotein(a). J. biol. Chem. *259:* 11470–11478 (1984).

17 Florentin, R. A.; Nam, S. C.; Daoud, A. S.; Jones, R.; Scott, R. F.; Morrison, E. S.; Kim, D. N.; Lee, K. T.; Thomas, W. A.; Dodd, W. J.; Miller, K. D.: Dietary-induced atherosclerosis in miniature swine. Parts I–V. Exp. molec. Path. *8:* 263–301 (1968).

18 Flow, B. L.; Cartwright, T. C.; Kuehl, T. J.; Mott, G. E.; Draemer, D. C.; Kruski, A. W.; Williams, J. D.; McGill, H. C., Jr.: Genetic effects on serum cholesterol concentrations in baboons. J. Hered. *72:* 97–103 (1981).

19 French, J. E.; Jennings, M. A.: The tunica intima of arteries of swine; in Roberts, Comparative atherosclerosis, pp. 25–36 (Harper & Row, New York 1965).

20 French, J. E.; Jennings, M. A.; Poole, J. C. F.; Robinson, D. S.; Sir Howard Florey: Intimal changes in the arteries of ageing swine. Proc. R. Soc. B *158:* 24–42 (1963).

21 Fuster, V.; Lie, J. T.; Badimon, L.; Rosemark, J. A.; Badimon, J. J.; Bowie, W.:

Spontaneous and diet-induced coronary atherosclerosis in normal swine and swine with von Willebrand disease. Arteriosclerosis 5: 67–73 (1985).

22 Gerrity, R. G.: The role of the monocyte in atherogenesis. I. Transition of blood-borne monocytes into foam cells in fatty lesions. Am. J. Path. 103: 181–190 (1981).

23 Gerrity, R. G.; Goss, J. A.; Soby, L.: Control of monocyte recruitment by chemotactic factor(s) in lesion-prone areas of swine aorta. Arteriosclerosis 5: 55–66 (1985).

24 Gerrity, R. G.; Naito, H. K.; Richardson, M.; Schwartz, C. J.: Dietary induced atherogenesis in swine. Morphology of the intima in prelesion stages. Am. J. Path. 95: 775–792 (1979).

25 Getty, R.: The gross and microscopic occurrence and distribution of spontaneous atherosclerosis in the arteries of swine; in Roberts, Comparative atherosclerosis, pp. 11–20 (Harper & Row, New York 1965).

26 Glomset, J. A.: The plasma lecithin-cholesterol acyltransferase reaction. J. Lipid Res. 9: 155–167 (1968).

27 Gofman, J. W.; Lindgren, F.; Elliot, H.; Mantz, W.: Hewitt, J.; Strisower, B.; Herring, V.: The role of lipids and lipoproteins in atherosclerosis. Science 3: 166–171, 186 (1950).

28 Goldsmith, D. P. J.; Jacobi, H. P.: Atherogenesis in swine fed several types of lipid-cholesterol diets. Lipids 13: 174–189 (1978).

29 Gottlieb, H.; Lalich, J. J.: The occurrence of arteriosclerosis in the aorta of swine. Am. J. Path. 30: 851–855 (1954).

30 Greer, S. A. N.; Hays, V. W.; Speer, V. C.; McCall, J. T.: Effect of dietary fat, protein and cholesterol on atherosclerosis in swine. J. Nutr. 90: 183–190 (1966).

31 Griggs, T. R.; Bauman, R. W.; Reddick, R. L.; Read, M. S.; Koch, G. G.; Lamb, M. A.: Development of coronary atherosclerosis in swine with severe hypercholesterolemia. Arteriosclerosis 6: 155–165 (1986).

32 Hamsten, A.; Iselius, L.; Dahlen, G.; Faire, U. de: Genetic and cultural inheritance of serum lipids, low and high density lipoprotein cholesterol and serum apolipoproteins A-I, A-II and B. Atherosclerosis 60: 199–208 (1986).

33 Hasler-Rapacz, J.; Rapacz, J.: Lipoprotein immunogenetics in primates. I. Two serum B-lipoprotein allotypes (Lmbl and Lmbll) in rhesus monkeys and the LP-B immunological relationship with other primates. J. med. Primatol. 11: 352–379 (1982).

34 Havel, R. J.; Eder, H. A.; Bragdon, J. H.: The distribution and chemical composition of ultracentrifugally separated lipoproteins in human serum. J. clin. Invest. 34: 1345–1353 (1955).

35 Hill, E. G.; Lundberg, W. O.; Titus, J. L.: Experimental atherosclerosis. I. A comparison of menhaden-oil supplements in tallow and coconut-oil diets. Mayo Clin. Proc. 46: 613–620 (1971).

36 Hill, E. G.; Lundberg, W. O.; Titus, J. L.: Experimental atherosclerosis in swine. II. Effects of methionine and menhaden-oil on an atherogenic diet containing tallow and cholesterol. Mayo Clin. Proc. 46: 621–625 (1971).

37 Hoff, H. F.; Gerrity, R. G.; Naito, H. K.; Dusek, D. M.: Methods in laboratory investigations. Quantitation of apo-B in aortas of hypercholesterolemic swine. Lab. Invest. 48: 492–504 (1983).

38 Iselius, L.; Carlson, L. A.; Morton, N. E.; Efendic, S.; Lindsten, J.; Luft, R.: Genetic and environmental determinants for lipoprotein concentrations in blood. Acta med. scand. 217: 161–170 (1985).

39 Jacobsson, L.: Comparison of experimental hypercholesterolemia and atherosclerosis in Göttingen mini-pigs and Swedish domestic swine. Atherosclerosis 59: 205–213 (1986).

40 Jackson, R. L.: Plasma lipoproteins of swine: composition, structure and metabolism; in Lewis, CRC handbook of electrophoresis, vol. IV (1983).

41 Jackson, R. L.; Baker, H. N.; Taunton, O. D.; Smith, L. C.; Garner, C. W.; Gotto, A. M., Jr.: A comparison of the major apolipoprotein from pig and human high density lipoproteins. J. biol. Chem. *248:* 2639–2644 (1973).

42 Jackson, R. L.; Taunton, O. D.; Segura, R.; Gallagher, J. G.; Hoff, H. F.; Gotto, A. M.: Comparative studies on plasma low density lipoproteins from pig and man. Compar. Biochem. Physiol. *53B:* 245–253 (1976).

43 Janado, M.; Martin, W. G.: Molecular heterogeneity of a pig serum low-density lipoprotein. Can. J. Biochem. *46:* 875–879 (1968).

44 Janado, M.; Martin, W. G.; Cook, W. H.: Separation and properties of pig-serum lipoproteins. Can. J. Biochem. *44:* 1201–1209 (1966).

45 Jennings, M. A.; Florey, H. W.; Stehbens, W. E.; French, J. E.: Intimal changes in the arteries of a pig. J. Path. Bact. *81:* 49–61 (1961).

46 Kalab, M.; Martin, W. G.: Gel filtration of native and modified pig serum lipoproteins. J. Chromat. *35:* 230–233 (1968).

47 Khan, M. A.; Earl, F. L.; Farber, T. M.; Miller, E.; Husain, M. M.; Nelson, E.; Gertz, S. D.; Forbes, M. S.; Rennels, M. L.; Heald, F. P.: Elevation of serum cholesterol and increased fatty streaking in egg yolk-lard fed castrated miniature pigs. Exp. molec. Pathol. *26:* 63–74 (1977).

48 Kim, D. N.; Imai, H.; Schmee, J.; Lee, K. T.; Thomas, W. A.: Intimal cell mass-derived atherosclerotic lesions in the abdominal aorta of hyperlipidemic swine. Atherosclerosis *56:* 169–188 (1985).

49 Knipping, G. M. J.; Kostner, G. M.; Holasek, A.: Studies on the composition of pig serum lipoproteins – isolation and characterization of different apoproteins. Biochim. biophys. Acta *393:* 88–99 (1975).

50 Lacko, A. G.; Lee, S-M.; Mirshahi, I.; Hasler-Rapacz, J.; Rapacz, J.: Reduced LCAT activity in hypercholesterolemic pigs with apolipoprotein mutations. Atherosclerosis (submitted).

51 Lee, D. M.; Mok, T.; Hasler-Rapacz, J.; Rapacz, J.: Lipid and lipoprotein profiles in pigs with mutant apolipoprotein-B and hypercholesterolemia, homozygous and heterozygous. Fed. Proc. *46:* 1340 (1987).

52 Lee, D. M.; Mok, T.; Hasler-Rapacz, J.; Rapacz, J.: Chemical composition of lipoprotein subfractions in swine with an apolipoprotein B mutation associated with hypercholesterolemia (Abstract). Arteriosclerosis *7:* 493 (1987).

53 Lee, K. T.: Cardiovascular; in Tumbleson, Swine in biomedical research, vol. 3, pp. 1481–1672 (Plenum Press, New York 1986).

54 Lee, K. T.; Jarmolych, J.; Kim, D. N.; Grant, C.; Krasney, J. A.; Thomas, W. A.; Bruno, A. M.: Production of advanced coronary atherosclerosis, myocardial infarction and 'sudden death' in swine. Exp. molec. Pathol. *15:* 170–190 (1971).

55 Lee, W. M.; Lee, K. T.; Thomas, W. A.: Partial suppression by pyridinolcarbamate of growth and necrosis of atherosclerotic lesions in swine subjected to an atherogenic regimen that produces advanced lesions. Exp. molec. Pathol. *30:* 85–93 (1979).

56 Instituto Graphico de Agostini: Leonardo da Vinci, p. 383 (Reynal, New York 1956).

57 Levy, R. I.; Stone, N. J.: Atherosclerosis: Role of lipoproteins; in Wissler, The pathogenesis of atherosclerosis, pp. 227–238 (Williams & Wilkins, Baltimore 1972).

58 Lewis, L. A.; Page, I. H.: Hereditary obesity: relation to serum lipoproteins and protein concentrations in swine. Circulation *14:* 55–59 (1956).

59 Luginbühl, H.: Spontaneous atherosclerosis; in Bustad, Swine in biomedical research, pp. 347–363 (Frayn, Seattle 1966).

60 Luginbühl, H.; Ratcliffee, H.L.; Detweiler, D.K.: Failure of egg-yolk feeding to accelerate progress of atherosclerosis in older femal swine. Virchows Arch. Abt. A Path. Anat. *348:* 281–289 (1969).

61 Maeda, N.; Ebert, D.L.; Doers, T.M.; Newman, M.; Hasler-Rapacz, J.; Attie, A.D.; Rapacz, J.; Smithies, O.: Molecular genetics of the apolipoprotein B gene in pigs in relation to atherosclerosis. Gene *70:* 213–229 (1988).

62 Mahley, R.W.: Dietary fat, cholesterol and accelerated atherosclerosis; in Paoletti, Atherosclerosis reviews, vol. 5, pp. 1–34 (Raven Press, New York 1979).

63 Mahley, R.W.: Cholesterol-induced hyperlipoproteinemia and atherosclerosis in dogs, swine and monkeys: Models for human atherosclerosis; in Gotto, Proc. 5th Int. Symp. on Atherosclerosis, pp. 355–358 (Springer, New York 1979).

64 Mahley, R.W.: Atherogenic lipoproteins and coronary artery disease: concepts derived from recent advances in cellular and molecular biology. Circulation *72:* 943–948 (1985).

65 Mahley, R.W.; Innerarity, T.L.; Rall, S.C.; Weisgraber, K.H.: Plasma lipoproteins: apolipoprotein structure and function. J. Lipid Res. *25:* 1277–1293 (1984).

66 Mahley, R.W.; Weisgraber, K.H.: An electrophoretic method for the quantitative isolation of human and swine lipoproteins. Biochemistry *13:* 1964–1969 (1974).

67 Mahley, R.W.; Weisgraber, K.H.; Innerarity, T.; Brewer, H.B., Jr.; Assmann, G.: Swine lipoproteins and atherosclerosis. Changes in the plasma lipoproteins and apoproteins induced by cholesterol feeding. Biochemistry *14:* 2817–2823 (1975).

68 Mancini, G.; Carbonara, A.O.; Heremans, J.F.: Immunochemical quantitation of antigens by single radial immunodiffusion. Int. J. Immunochem. *2:* 235–254 (1965).

69 McConathy, W.J.; Alaupovic, P.: Isolation and partial characterization of apolipoprotein D: a new protein moiety of the human plasma lipoprotein system. FEBS Lett. *37:* 178–182 (1973).

70 Moreland, A.F.; Clarkson, T.B.; Lofland, H.B.: Atherosclerosis in 'Miniature' swine. I. Morphologic aspects. Archs Path. *76:* 203–210 (1963).

71 Pond, W.G.; Mersmann, H.J.; Young, L.D.: Heritability of plasma cholesterol and triglyceride concentrations in swine. Proc Soc. exp. Biol. Med. *182:* 221–224 (1986).

72 Pownall, H.J.; Jackson, R.L.; Roth, R.I.; Gotto, A.M.; Patsch, J.R.; Kummerow, F.A.: Influence of an atherogenic diet on the structure of swine low density lipoproteins. J. Lipid Res. *21:* 1108–1115 (1980).

73 Rapacz, J.: Immunogenetic polymorphism and genetic control of low density β-lipoproteins in swine. Proc. 1st Wld Congr. on Genetics Applied to Livestock Production, Madrid, vol. I, pp. 291–298 (Graficas Orbe, Madrid 1974).

74 Rapacz, J.: Lipoprotein immunogenetics and atherosclerosis. Am. J. med. Genet. *1:* 377–405 (1978).

75 Rapacz, J.: Current status of lipoprotein genetics applied to livestock production in swine and other domestic species. Proc. 2nd Wld Congr. on Genetics Applied to Livestock Production, Madrid, vol. VI, pp. 365–374 (Graficas Orbe, Madrid 1982).

76 Rapacz, J.; Elson, C.E.; Lalich, J.J.: Correlation of an immunogenetically defined lipoprotein type with aortic intimal lipidosis in swine. Exp. molec. Pathol. *27:* 249–261 (1977).

77 Rapacz, J.; Grummer, R.H.; Hasler, J.; Shackelford, R.M.: Allotype polymorphism of low density β-lipoproteins in pig serum (LDLppl and LDLpp2). Nature, Lond. *225:* 941–942 (1970).

78 Rapacz, J.; Hasler-Rapacz, J.: Investigations in the relationship between immunogenetic polymorphisms of β-lipoproteins and the β-lipoprotein and cholesterol

levels in swine; in Lenzi, Atherosclerosis and cardiovascular diseases, pp. 99–108 (Editrice Compositori, Bologna 1984).

79 Rapacz, J.; Hasler, J.; Duniec, M.; Kazana, J.: Serum antigens of β-lipoprotein in pigs (LDLpp3); in Kovacz, Proc. 12th Eur. Conf. on Animal Blood Groups and Biochemical Polymorphisms, pp. 383–385 (Junk, The Hague 1972).

80 Rapacz, J.; Hasler-Rapacz, J.; Kuo, W. H.: Immunogenetic polymorphism of lipoproteins in swine. 2. Five new allotypic specificities (Lpp6, Lpp11, Lpp12, Lpp13 and Lpp14) in the Lpp system. Immunogenetics 6: 405–424 (1978).

81 Rapacz, J.; Hasler-Rapacz, J.; Kuo, W. H.: Immunogenetic polymorphism of lipoproteins in swine: genetic, immunological and physiochemical characterization of two allotypes Lpr1 and Lpr2. Genetics 113: 985–1007 (1986).

82 Rapacz, J.; Hasler-Rapacz, J.; Kuo, W. H.; Li, D.: Immunogenetic polymorphism of lipoproteins in swine. I. Four additional serum lipoprotein allotypes (Lpp2, Lpp4, Lpp5 and Lpp15) in the Lpp system. Anim. Blood Groups Biochem. Genet. 7: 157–177 (1976).

83 Rapacz, J.; Hasler-Rapacz, J.; Taylor, K. M.; Checovich, W. J.; Attie, A. D.: Immunogenetically defined lipoprotein polymorphism associated with hypercholesterolemia and accelerated atherosclerosis in swine (Abstrat). 6th Int. Meet. on Atherosclerosis and Cardiovascular Diseases, Bologna 1986, p.318.

84 Rapacz, J.; Hasler-Rapacz, J.; Taylor, K. M.; Checovich, W. J.; Attie, A. D.: Lipoprotein mutations in pigs are associated with elevated plasma cholesterol and atherosclerosis. Science 234: 1573–1577 (1986).

85 Rapacz, J., Jr.; Hasler-Rapacz, J.; Rapacz, J.; McConathy, W. J.: Separation of swine plasma LDL from Apo-B heterozygotes into two allelic products, Lpb2 and Lpb3 (Abstract). Arteriosclerosis 7: 493 (1987).

86 Rapacz, J., Jr.; Hasler-Rapacz, J.; Rapacz, J.; McConathy, W. J.: Separation of swine plasma LDL from Lpb2/3 heterozygotes into two apo-B allelic haplotypes, Lpb2 and Lpb3, with apo-B epitope specific antibodies. J. Lipid Res. (in press).

87 Rapacz, J., Jr.; Reiner, Z.; Ye, S. Q.; Hasler-Rapacz, J.; Rapacz, J.; McConathy, W. J.: Plasminogen polymorphism in swine. Comp. Biochem. Physiol. (in press).

88 Ratcliffe, H. L.; Luginbuhl, H.; Pivnik, L.: Coronary, aortic and cerebral atherosclerosis in swine of 3 age-groups: implications. Bull. Wld Hlth Org. 42: 225–234 (1970).

89 Reiser, R.; Sorrels, M. F.; Williams, M. C.: Influence of high levels of dietary fats and cholesterol on atherosclerosis and lipid distribution in swine. Circulation Res. 7: 833–846 (1959).

90 Reitman, J. S.; Mahley, R. W.: Changes induced in the lipoproteins of Yucatan miniature swine by cholesterol feeding. Biochim. biophys. Acta 575: 446–457 (1979).

91 Reitman, J. S.; Mahley, R. W.; Fry, D. L.: Yucatan miniature swine as a model for diet-induced atherosclerosis. Atherosclerosis 43: 119–132 (1982).

92 Roberts, J. C., Jr.; Stras, R.: Comparative atherosclerosis (Harper & Row, New York 1965).

93 Rosenfeld, M. E.; Prescott, M. F.; McBride, C.; Rapacz, J.; Hasler-Rapacz, J.; Attie, A. D.: The composition and distribution of atherosclerotic lesions in the Lpb-5 spontaneously hypercholesterolemic pig closely resemble human atherosclerotic lesions (Abstract). FASEB – Pathology (1988).

94 Rothschild, M. F.; Chapman, A. B.: Factors influencing serum cholesterol levels in swine. J. Hered. 67: 47–48 (1976).

95 Rowsell, H. C.; Downie, H. G.; Mustard, J. F.: The experimental production of atherosclerosis in swine following the feeding of butter and margarine. Can. med. Ass. J. 79: 647–654 (1958).

96 Rowsell, H. C.; Downie, H. G.; Mustard, J. F.: Comparison of the effect of egg yolk or butter on the development of atherosclerosis in swine. Can. med. Ass. J. *83:* 1175–1186 (1960).

97 Rowsell, H. C.; Mustard, J. F.; Downie, H. G.: Experimental atherosclerosis in swine. Ann. N. Y. Acad. Sci. *127:* 742–762 (1965).

98 Scott, R. F.; Daoud, A. S.; Florentin, R. A.: Animal models in atherosclerosis; in Wissler, The pathogenesis of atherosclerosis, pp. 120–146 (Williams & Wilkins, Baltimore 1972).

99 Scott, R. F.; Kim, D. N.; Schmee, J.; Thomas, W. A.: Atherosclerotic lesions in coronary arteries of hyperlipidemic swine. Atherosclerosis *62:* 1–10 (1986).

100 Scott, R. F.; Reidy, M. A.; Kim, D. N.; Schmee, J.; Thomas, W. A.: Intimal cell mass derived atherosclerotic lesions in the abdominal aorta of hyperlipidemic swine. Atherosclerosis *62:* 27–38 (1986).

101 Segrest, J. P.; Albers, J. J.: Plasma lipoproteins. Part A. Meth. Enzym. *128* (1986).

102 Skold, B. H.; Getty, R.: Spontaneous atherosclerosis in swine. J. Am. vet. med. Ass. *139:* 655–660 (1961).

103 St. Clair, R. W.; Bullock, B. C.; Lehner, N. D. M.; Clarkson, T. B.; Loftland, H. B., Jr.: Long-term effects of dietary sucrose and starch on serum lipids and atherosclerosis in miniature swine. Exp. molec. Pathol. *15:* 21–33 (1971).

104 Stunzi, H.: Informal notes on arterial lesions in swine. Ann. N. Y. Acad. Sci. *127:* 740–742 (1965).

105 Tall, A. R.; Atkinson, D.; Small, D. M.; Mahley, R. W.: Characterization of the lipoproteins of atherosclerotic swine. J. biol. Chem. *252:* 7288–7293 (1977).

106 Task Force on Genetic Factors in Atherosclerotic Disease, 1974/1975, Washington, DHEW No. NIH 76–922.

107 Thomas, W. A.; Kim, D. N.; Lee, K. T.; Reiner, J. M.; Scheme, J.: Population of dynamics of arterial cells during atherogenesis. Exp. molec. Pathol. *39:* 257–270 (1983).

108 Watanabee, Y.: Serial inbreeding of rabbits with hereditary hyperlipemia (WHHL-rabbit). Incidence and development of atherosclerosis. Atherosclerosis *36:* 261–268 (1980).

109 Wissler, R. W.; Vesselinovitch, D.: Differences between human and animal atherosclerosis; in Schettler, Proc. 3rd Int. Symp. on Atherosclerosis, 1973 (Springer, New York 1974).

Dr. Jan Rapacz, University of Wisconsin Immunogenetics Laboratory, 666 Animal Sciences Building, Madison, WI 53706 (USA)

Lusis A J, Sparkes S R (eds): Genetic Factors in Atherosclerosis: Approaches and Model Systems. Monogr Hum Genet. Basel, Karger, 1989, vol 12, pp 170–188

Genetic Aspects of Plasma Lipoprotein and Cholesterol Metabolism in Nonhuman Primate Models of Atherosclerosis

Kathy Laber-Laird, Lawrence L. Rudel

Arteriosclerosis Research Center, Departments of Comparative Medicine, and Biochemistry, The Bowman Gray School of Medicine of Wake Forest University, Winston-Salem, N. C., USA

In human beings and in experimental animals, dietary cholesterol and saturated fat induce increases in plasma lipoprotein cholesterol and apolipoprotein concentrations [1–3]. However, the mechanisms regulating the relationships between dietary cholesterol, plasma cholesterol and tissue cholesterol metabolism are not well understood [4]. A better understanding of these mechanisms is needed because plasma lipoprotein concentrations are highly correlated with coronary heart disease and the extent of coronary artery atherosclerosis can be decreased by lowering plasma and LDL cholesterol concentrations [2, 5, 6].

A high degree of individual variability occurs in the response to dietary cholesterol and fat. This variability is thought to be due to a combination of environmental and genetic influences, but the relative contribution of each on cholesterol metabolism is difficult to quantitate [7]. As a general rule, phenotypic correlations between lipoprotein endpoints and physiologic processes known to influence cholesterol metabolism are low. However, these correlations are probably low because a major component of the variation is random from uncontrolled or unacknowledged variation. A more thorough understanding of the genetic influences on plasma lipoprotein cholesterol concentrations could help explain individual differences in cholesterol metabolism and open avenues for prevention of coronary heart disease.

Several species of nonhuman primates have been established as models for the study of lipoprotein metabolism and atherosclerosis [8–11]. Lipoprotein compositions and distributions generally resemble those of human beings when similar high fat diets are fed [9] and significant correlations of atherosclerosis extent with LDL and HDL cholesterol concentrations have been reported in several species [12–15]. Primates are phylogenetically close to man which is an advantage when so many variables affecting lipoprotein metabolism and atherosclerosis are not defined and therefore unable to be controlled experimentally. In addition, nonhuman primates are one of the few animal models that have a significant degree of CNS control over physiologic events [16]. Although among primates, the great apes appear to share more plasma lipoprotein similarities with man than monkeys, their lack of availability limits their usefulness as experimental models. On the other hand, several species of monkeys are readily available for research. For these reasons, monkeys are the most practical and valuable primate models for studying genetic and environmental influences on lipoprotein and cholesterol metabolism and atherosclerosis. The biggest drawbacks to the use of these primates as genetic models are: (1) the expense and complications of maintenance, and (2) the long gestation time and propensity to single births, factors which make genetic studies in monkeys infrequent and difficult.

Characteristic Lipoprotein Cholesterol Distribution in Squirrel Monkeys (Saimiri sciureus)

Some of the initial studies using primate animal models to look at the genetic influences on plasma cholesterol were done with a New World primate, the squirrel monkey [17–21]. Male squirrel monkeys eating a native diet had a mean total plasma cholesterol concentration (TPC) of 103 mg/dl with 31% of the cholesterol in HDL, and females eating the same diet had an average TPC of 106 mg/dl with 27% of the cholesterol in HDL [17]. In a mixed population of captive monkeys fed monkey chow, the average TPC was 182 mg/dl with 42% located in HDL. The higher total cholesterol was thought to be due to the stress of capture and relocation [18]. Rudel and Lofland [8] have documented that squirrel monkeys respond to a cholesterol-enriched diet (0.5% cholesterol) with plasma cholesterol concentrations of 378 mg/dl with 5% cholesterol in VLDL, 7% in intermediate-sized LDL (ILDL), 50% in LDL, and 38% in HDL. The greatest percentage increase induced by dietary cholesterol was in the cholesterol concentration of VLDL while the highest absolute increase in cholesterol concentration occurred in LDL. HDL-cholesterol was slightly increased.

Genetic Influences on Plasma Cholesterol in Squirrel Monkeys

In 1971, Clarkson et al. [19] identified two groups of squirrel monkeys that had either an over- or underexaggerated response to dietary cholesterol when compared to the majority of monkeys fed the same diet. The two groups of animals were termed hyperresponders and hypo-responders, respectively. The hyperresponders developed severe hyper-cholesterolemia (770–1,000 mg/dl) while the hyporesponders maintained plasma cholesterol concentrations similar to animals fed a control diet (170–250 mg/dl). Animals from the two groups were selectively bred and the responsiveness of the offspring to various concentrations of dietary cholesterol was determined. It was noted that prior to feeding cholesterol, differences between progeny could be detected in TPC concentrations. The animals sired from hyporesponders had an average plasma cholesterol of 270 mg/dl while those sired from hyperresponders had an average plasma cholesterol of 369 mg/dl. The differences between groups became much more evident when cholesterol was fed. After 3 months of eating a semipurified diet containing 0.5% cholesterol, the progeny of hyperresponders had a mean serum cholesterol of 1,030 mg/dl while those of hyporesponders had an average serum cholesterol of 311 mg/dl.

A numerical estimate of heritability was determined by comparing the regression of the plasma cholesterol values of the offspring on the midparental mean [19]. The heritability estimate was 0.46. The offspring were also ranked based on their plasma cholesterol response and the non-parametric correlation coefficient was 0.75. The statistical significance of the rank correlation helped to confirm that a genetic influence on plasma cholesterol concentrations occurred in response to diet. Atherosclerosis in the hyperresponsive animals was more extensive and severe than typi-cally seen in this species. The hyperresponders also had a more frequent incidence of cutaneous xanthomatosis and lesions associated with hyper-lipidemia in other organs. It was suggested that although the genetic influences were only determined on plasma cholesterol, the severity of atherosclerosis may also be heritable and related in some way to the genetic influence on TPC.

Evidence for Genetic Influence on Cholesterol Metabolism in Squirrel Monkeys

In 1972, Lofland et al. [20] investigated several potential sites of dif-ference in regulation of plasma cholesterol between the two groups of squirrel monkeys. Cholesterol absorption and synthesis were found to be similar between the two groups but hyporesponders had increased bile acid excretion sooner after dietary cholesterol was fed and to a greater degree than hyperresponders. Hyporesponders had faster cholesterol turnover rates and smaller body cholesterol pools than did hyperresponders. Neutral sterol excretion was similar between groups. This suggested that, in squirrel monkeys, there may have been a genetic effect on the mechanisms regulating bile acid excretion that was sensitive to cholesterol influx when dietary cholesterol was fed [20]. In a later study by Jones et al. [21], sterol balance techniques indicated that hyporesponsive squirrel monkeys absorbed less dietary cholesterol than hyperresponsive monkeys. The difference in this outcome from the earlier result is not clear but may be related to small sample size, differences in diet, or other unknown factors. In other studies using these two groups of squirrel monkeys, Guertler and St. Clair [22] found that the hyperresponsiveness was not associated with an inability to down-regulate the LDL receptor, as demonstrated in incubations of skin fibroblasts.

Characteristic Lipoprotein Distributions in Rhesus Monkeys (Macaca mulatta)

Rhesus monkeys are a species of Old World monkey in which the lipoprotein response to dietary cholesterol has been characterized extensively. In male rhesus monkeys fed a monkey chow diet, Lee and Morris [23] reported the average TPC to be 165 mg/dl with 1% in VLDL, 1% in LDL_1, 37% in LDL_2, and 60% in HDL. Other studies by Rudel et al. [24] have also shown that in normal male rhesus monkeys fed high fat, low cholesterol diets, the lipoprotein cholesterol distribution is similar to that shown for animals fed monkey chow, i.e. TPC, 190; VLDL, 1%; ILDL, 14%, LDL, 30%, and HDL, 55%. In this study, it was also identified that the rhesus monkey as a species, and some individuals, in particular, have high concentrations of Lp(a), a pre-β-migrating lipoprotein with a density interval between 1.05 and 1.1 g/ml. This result confirmed a similar finding of Nelson and Morris [25]. In this species more than in others, because of the high Lp(a) concentration, the separation of lipoproteins according to the classical density intervals (1.006, 1.019, 1.063, and 1.21 g/ml) for VLDL, IDL, LDL, and HDL did not result in purified lipoprotein classes [24]. In response to a high fat diet containing a higher concentration of cholesterol (0.5% (w/w) or 1 mg/kcal), the average serum cholesterol response for 6 animals was 822 mg/dl with 6% of plasma cholesterol in VLDL, 11% in ILDL, 79% in LDL and 4% in HDL [26]. The greatest increase in cholesterol concentration occurred in the LDL fraction, which also increased dramatically in particle size and cholesteryl ester content. Significant increases also were found in the concentrations of ILDL and

VLDL and a significant amount of β-VLDL was identified. The concentration of the Lp(a) lipoproteins did not increase with dietary cholesterol challenge, and the HDL concentrations actually decreased in these rhesus males fed diets enriched in cholesterol.

A strong relationship between the extent of atherosclerosis and the degree of diet-induced hypercholesterolemia has been demonstrated in rhesus monkeys by Armstrong et al. [13]. These workers showed that only modest amounts of dietary cholesterol (0.129 mg/kcal) induced small increases in plasma cholesterol concentrations ($\sim$50 mg/dl) and higher LDL/HDL ratios. These changes occurred together with an increased extent of atherosclerosis. An extensive study of atherosclerosis regression in rhesus monkeys has also shown that when plasma cholesterol concentrations were reduced from 350 to 250 mg/dl, more coronary artery atherosclerosis regression occurred [27]. Taken together, these data demonstrate that the rhesus monkey is a valuable model for development of pertinent information about associated environmental and genetic mechanisms affecting plasma lipoproteins and atherosclerosis.

Estimates of heritability for plasma cholesterol response have not been reported in the literature for this species, but there are reports of natural and diet-induced hyperlipoproteinemias which appear to be genetically mediated.

Evidence for Genetic Influence on Cholesterol Metabolism in Rhesus Monkeys

In 1976, Eggen [28] identified a population of rhesus monkeys with a heterogeneous response to dietary cholesterol. The population was divided into high and low responders and differences in cholesterol metabolism were compared between the groups. The data suggested that lumenal cholesterol absorption was greater in the high responders than in the low responders, and the excretion of bile acids was also apparently greater in the high responders. This outcome was different from that obtained in squirrel monkeys as discussed above [20]. This may be a reflection of differences in experimental conditions, or a reflection of different mechanisms operating in different species. Only the squirrel monkeys were genetically selected before study, although even in this case the degree of selection was uncertain since numerous breeders were used and first-generation progeny were studied. Clearly, several genetic factors are likely to be contributing to the outcome in either experiment.

Genetic Hyperlipoproteinemias in Rhesus Monkeys

In 1968, Morris and Fitch [29] reported a naturally occurring hyper-betalipoproteinemia in two adult male rhesus monkeys. The average plasma cholesterol concentrations for these animals was significantly higher (472 vs. 144 mg/dl) compared to those of other rhesus monkeys fed an identical diet. These monkeys carried 21% of their cholesterol in VLDL, 22% in IDL, 54% in LDL, and 3% in HDL. Much more of the serum cholesterol was carried in VLDL and markedly less was found in HDL compared to the other monkeys studied. This cholesterol distribution was more similar to that noted by Rudel et al. [26] in rhesus monkeys fed dietary cholesterol (see above) than that for normal rhesus monkeys; however, the LDL in the spontaneously hypercholesterolemic monkeys was of normal size and composition, and was not an abnormally large LDL as in the diet-induction studies [23, 30]. Therefore, this lipoprotein abnormality appeared phenotypically similar to the hyperbetalipoproteinemia of familial hypercholesterolemia (type IIB hyperlipoproteinemia) in human beings [31].

Data from these two male rhesus monkeys suggested that the defect is genetic, although the site of the genetic defect remains unidentified. Although no females with a similar abnormality were available as breeders, one male progeny from one of the two hyperbetalipoproteinemic males had hypercholesterolemia [32]. When several progeny from both males were compared to progeny from matings of normal animals, their plasma cholesterol responses to dietary cholesterol were significantly elevated [33]. Interestingly, all of the progeny from one male had an elevated plasma cholesterol response to the dietary challenge, while only half of the progeny from the other male were abnormal. In isolated skin fibroblasts from the two index male monkeys, the LDL receptor binding, uptake, and metabolism did not appear to be abnormal [34], so that these animals did not appear to be models of the classical LDL receptor deficiency described by Brown and Goldstein [35]. However, these results in skin fibroblasts, where all of the available LDL receptor is up-regulated, do not rule out the possibility that other abnormalities in LDL receptor function occur that may not have been clearly demonstrated in fibroblast culture. For example, in the skin fibroblast samples from one of the monkeys with spontaneous hyperbetalipoproteinemia, the regulation of HMG-CoA reductase by cholesterol internalized via the LDL receptor pathway was abnormally low. Further studies to define the defect in these monkeys are needed.

Morris and Greer [36] also identified a population of rhesus monkeys that responded with an increase rather than a decrease in HDL cholesterol concentration in response to dietary cholesterol. In these monkeys, the proportion of TPC as HDL cholesterol was greater than 40% at all times and their TPC were approximately 250 mg/dl less than monkeys that responded to dietary cholesterol with the more typical decrease in HDL cholesterol [36]. Rudel and Lofland [8] were also able to identify a subgroup of rhesus monkeys that had significantly lower serum cholesterol concentrations but significantly higher HDL-C concentrations in response to dietary cholesterol when compared to the group as a whole. These data indicate that control of HDL cholesterol concentration is related to plasma cholesterol response and is unique in a subset of the rhesus population. The latter observation implies that a genetic influence exists [8].

Characteristic Lipoprotein Distributions in Cynomolgus Monkeys (Macaca fascicularis)

Cynomolgus monkeys are a species of macaque used in atherosclerosis research with increasing frequency. As a species, these animals are among the most sensitive to dietary cholesterol and their plasma cholesterol response shows exaggerated characteristics in a number of lipoprotein endpoints. The data of Tall et al. [37] show that in a group of 36 male cynomolgus monkeys fed diets containing 0.3% (w/w) or 0.7 mg/kcal of cholesterol, the average TPC was 823 mg/dl. This value compares to a TPC of 164 mg/dl in a control group of animals fed the same diet, but with 0.015% cholesterol. The cholesterol concentrations (mg/dl) among lipoprotein fractions was 16 in VLDL, 105 in ILDL, 653 in LDL, and 22 in HDL. These values contrast to cholesterol concentrations of 2 in VLDL, 14 in ILDL, 80 in LDL and 68 in HDL in the control diet-fed animals.

The average size of the LDL particles also dramatically increases in response to dietary cholesterol [37], and the size increase occurs to a greater extent in males than in females [38]. This increase in LDL size has been shown to be due to an enrichment of the lipoprotein particles with cholesteryl esters, primarily cholesteryl oleate, in animals fed saturated fat in the diet. This selective enrichment with cholesteryl oleate suggests that the cholesteryl esters accumulating in the plasma LDL particles is derived from a tissue source of acyl-CoA: cholesterol acyltransferase. The tissue source is presumably the liver since the plasma LDL in this species and in other nonhuman primate species has been shown to contain only apo B_{100}, the form of apo B derived from the liver [39]. The shift in cholesteryl ester composition is associated with an increases in the transition temperature of the LDL such that the melting temperature of the cholesteryl ester core of the LDL is above body temperature in the cholesterol-fed cynomolgus monkey [37]. The size of the LDL in cynomolgus monkeys has been found to be highly correlated to the extent of coronary artery atherosclerosis [9], and it has been hypothesized that the physical state of the core cholesteryl esters may be contributing to the relative atherogenicity of these LDL. The metabolic pathway that leads to these changes in plasma LDL in cynomolgus monkeys is still under study. However, the fact that the degree of change that occurs is individual animal specific with a rank order among the animals that stays relatively constant with time, suggests genetic control of the process. Presumably such control is via a gene(s) regulating cholesterol metabolism in the liver, although this remains only speculative.

Evidence for Genetic Influences on Plasma Cholesterol in Cynomolgus Monkeys

Clarkson et al. [40] also have observed that significant variability in the TPC response to dietary cholesterol was present in cynomolgus monkeys. In 1977, selective breeding of these monkeys with extreme responses to dietary cholesterol was initiated. In response to 0.39 mg/kcal cholesterol, female monkeys classified as hyperresponders had an average cholesterol concentration of 480 ± 30 mg/dl with the male counterparts having an average of 544 ± 24 mg/dl. The hyporesponsive females and males responded with 228 ± 28 and 219 ± 56 mg/dl, respectively. Over 60 progeny have been produced from both sets of parents and it appears as though the progeny are responding to cholesterol in a manner similar to their parents although these data are still being evaluated [40]. This preliminary information suggests that the TPC response to dietary cholesterol exhibited by cynomolgus monkeys is also influenced by a genetic component, as appears to be the case in other species. Further documentation is needed.

Characteristic Lipoprotein Distribution in African Green Monkeys (Cercopithecus aethiops)

Although no direct estimates of heritability and genetic influences on plasma lipoproteins have been documented in African green monkeys, the literature on individual differences in lipoprotein metabolism in this species strongly supports the likelihood of genetic influences on cholesterol metabolism. Since the response to atherogenic diets resembles the response of human beings, these animals could become a model of genetic influences relevant to man.

TPC concentrations in African green monkeys respond to dietary cholesterol within a range more comparable to human beings than that of macaques and the lipoprotein distribution of these animals more closely mimics what has been documented in human beings with familial hyper-cholesterolemia [9]. In response to a cholestrol-enriched diet (0.4% cholesterol) the average TPC in a group of these animals increased from 144 to 399 mg/dl with 0.5% of the cholesterol present in VLDL, 7% in ILDL, 65% in LDL and 27% in HDL. The diet-induced increase in lipoprotein cholesterol occurred primarily in the LDL fraction although an increase in ILDL cholesterol also occurred. No significant increase in the average HDL or VLDL cholesterol concentration was observed, although a patterned response in HDL was observed among animals [9].

Evidence for Genetic Influences on Plasma Cholesterol and Cholesterol Metabolism in African Green Monkeys

St. Clair et al. [41] and Rudel et al. [42] have documented significant individual variability in the TPC response to diets enriched with cholesterol within populations and between subspecies of African green monkeys. This variability was associated with LDL and HDL cholesterol concentrations. In animals that responded to dietary cholesterol with high plasma cholesterol concentrations (>235 mg/dl), HDL cholesterol concentrations had an inverse relationship with TPC [42]. Some animals fed the same diet and animals fed a diet with a low cholesterol content had TPC values below 235 mg/dl; a significant positive relationship between HDL cholesterol and TPC existed for these animals. This suggests, by analogy with the observations in rhesus monkeys, that the HDL response is under genetic influence and the gene(s) that regulates HDL concentrations may also regulate TPC or some aspect of cholesterol metabolism that affects both HDL and TPC. Recent data from our laboratory [43] have demonstrated that the rate of hepatic apolipoprotein A-I production and the abundance of hepatic and intestinal apo A-I mRNA are much higher in African green monkeys than in cynomolgus monkeys. Plasma HDL concentrations in cynomolgus monkeys are lower, and these macaques are unable to increase HDL cholesterol concentrations in response to a diary cholesterol challenge, as has been seen in African green monkeys [Rudel, unpubl. observations]. Therefore, regulation of the apo A-I gene seems to be intimately related, albeit in an unknown fashion, to the trait of dietary cholesterol sensitivity. Interestingly, this property seems to be shared among several mammalian species [43].

Rudel et al. [42] compared the lipoprotein response to different diets in subspecies of African green monkeys (grivets vs. vervets). Average values for both TPC and HDL cholesterol concentrations were significantly higher in the grivet subspecies compared to the vervet subspecies. However, data from both subspecies could be fit to the same regression line correlating the two variables. This suggested that the difference between subspecies was due to magnitude of response instead of the mechanism of response [42]. If this is true, the phenotypic expression of the genetic differences between the subspecies would be related to the degree of genetic regulation of the same aspects of cholesterol metabolism.

The possibility that differences in cholesterol absorption or excretion explain individual variations among African green monkeys in response to dietary cholesterol also has been studied [41, 44]. However, in these experiments, no correlations between TPC and cholesterol absorption, excretion, or synthesis were observed. Apparently, the relationships between these parameters and TPC are controlled by several factors instead of any one of these parameters.

Characteristic Lipoprotein Distributions in Baboons
(Papio cynocephalus)

Probably the most extensive evaluation of the genetic influences on cholesterol metabolism in nonhuman primates has been done in the baboon. The baboon is an animal with low lipid and lipoprotein concentrations on baseline diets. De La Pena et al. [45] have studied 15 control male and 15 control female baboons. The average TPC in the females was 129 mg/dl and it was 111 mg/dl in the males. The percent distribution of cholesterol among lipoproteins was: for females, VLDL – 3%, IDL – 14%, LDL – 27%, and HDL – 56%; for males, VLDL – 3%, IDL – 5%, LDL – 27%, and HDL – 65%. The relatively higher amount of plasma cholesterol in IDL and LDL in females compared to males was significantly different. Plasma triglyceride concentrations were 29 mg/dl in females and 28 mg/dl in males. In general, the distributions of triglyceride were similar to those of cholesterol. This distribution of triglyceride, which is quite different from that seen in human beings, is more typical of many of the nonhuman primates, presumably because clearance mechanisms in these species are more efficient, as evidenced by the low concentrations.

Kushwaha et al. [46] fed four chair-restrained male baboons an atherogenic diet for 6 months. The average TPC increased from 98 to 202 mg/dl, with the largest portion of this increase occurring in LDL, the cholesterol concentration of which increased from 39 to 114 mg/dl. HDL cholesterol concentrations also increased from 41 to 68 mg/dl. The ratio of LDL cholesterol/HDL cholesterol was 0.95 before and 1.7 after the atherogenic diet.

McGill et al. [47] also studied the lipoproteins in baboons fed control and cholesterol-enriched diets. In addition to the types of changes just described, they measured apo B, apo A-I, and apo E concentrations and the diet response of these endpoints in baboons known to have high or low levels of HDL_1, an unusual high density lipoprotein frequently found in many baboons. The presence of low or high concentrations of HDL_1 has been shown to be a genetic trait [48]. Animals with high HDL_1 were generally animals with significantly higher concentrations of TPC, VLDL + IDL cholesterol, LDL cholesterol, apo A-I and apo E concentrations. In these carefully controlled experiments, several effects were evaluated including those of dietary cholesterol and the type of dietary fat, as well as the HDL_1 phenotype and the various interactions. The effects of dietary cholesterol were to significantly raise the concentrations of cholesterol and apo B in VLDL + IDL, LDL, and HDL_1. HDL_2 and HDL_3 cholesterol and apo A-I concentrations appeared to remain unchanged by dietary cholesterol. Based on this observation, it is possible that the increase in HDL cholesterol seen earlier in baboons by Kushwaha et al. [46] was due to the increase in HDL_1 lipoproteins.

Baboons also have relationships between atherosclerosis and lipoprotein concentrations that are consistent with findings from epidemiologic studies. There is a positive relationship between fatty streaks in the aorta, major aortic branches and coronary arteries with LDL/VLDL cholesterol while there is a negative relationship with HDL cholesterol concentrations [14].

Genetic Influences on Plasma and Lipoprotein Cholesterol Metabolism in Infant and Juvenile Baboons

Quantitative genetic techniques have been applied in this species in order to determine the genetic contribution that often is not apparent from phenotypic correlations. In 1973, an experiment was started at the Southwest Medical Foundation by Mott et al. [48] whose primary purpose was to evaluate the effects of breast and formula feeding on cholesterol metabolism in a group of 100 infants originating from 100 dams and 7 sires. Because the pedigree of each of the infants was known and the environment was well controlled, the genetic contribution to cholesterol metabolism could be determined. The first investigation evaluated the sire effect (the degree the offspring from a particular sire vary from the population mean) on TPC. There was a significant sire effect on TPC at birth as well as at 12 weeks of age. However, the sire effect at birth did not predict the effect at 12 weeks. The sire effect was not mediated through cholesterol absorption and there was trend (although not significant) toward a greater sire effect for high responders versus low responders. The conclusions were that (1) TPC concentrations are under partial genetic control but that this control is not related to cholesterol absorption, and (2) the biochemical mechanism of the genetic control varies with age [48].

From 12 weeks to 1 year of age, the rank order correlations of TPC among progeny and the rank order of sire progeny groups remained consistent. The heritability estimate ranged between 0.25 and 0.80 with the average being 0.45 by 1 year of age. These estimates are similar to what has been documented in other species including the squirrel monkey. The conclusion was made that the genes regulating serum cholesterol in infancy and adulthood were the same and the failure of values at birth to predict values in juveniles was probably a reflection of temporary environmental variance associated with in utero or maternal effects [49].

When the baboons reached 3.5 years of age, further investigations were carried out in order to assess the heritability of cholesterol metabolism. Flow and Mott [50] determined that the heritability for TPC had remained constant at the moderately high value of 0.44, and they were also able to determine that the whole-body cholesterol production rate was moderately to highly heritable (0.56) as was the cholesterol turnover rate (0.71). In contrast, the phenotypic correlations between TPC and cholesterol turnover and production rates were low (0.02, 0.24, respectively). They concluded that the low phenotypic correlation between TPC and cholesterol turnover was due to genetic and environmental contributions of similar but opposite magnitude. Speculation was that the environmental contribution included the effects of interaction (epistasis) with other physiologic factors such as individual variations in cholesterol absorption, for example. The diet was not thought to be a contributor because it was held constant. The low phenotypic correlation of TPC with cholesterol production rate was presumably due to the genetic contribution because the environmental contribution was minimal [50].

The genetic mediation of LDL cholesterol and apoprotein concentrations were determined in the same group of juvenile baboons (now aged 4–6 years) in 1982 by Flow et al. [51]. Significant differences were observed between sire progeny groups for serum and HDL cholesterol and apo A-I concentrations. A marginal sire effect was noted for VLDL/

LDL cholesterol. The estimates of heritability were high for HDL (0.78) moderate for apo A-I and TPC (0.56 and 0.54) and low for VLDL/LDL and apo B (0.32, 0.20, respectively). There was a phenotypic correlation of VLDL/LDL with HDL cholesterol (0.31) that was due entirely to a genetic contribution. A strong negative genetic correlation (−0.95) was documented between apo A-I and cholesterol turnover. The genetic correlation measures the extent that the two variables are influenced by the same gene or genes and the high genetic correlation between apo A-I and cholesterol turnover suggests a physiologic relationship between these two varibles. Apo B concentrations and cholesterol turnover rates were independent [51].

The heritability of TPC and apoproteins was similar to what has been documented in man; however, the phenotypic correlation between TPC and HDL cholesterol was much higher in baboons than in man. This may be related to the differences in HDL cholesterol concentrations and sub-fraction distributions between man and baboons. Human beings also have a low phenotypic correlation between LDL-C and HDL-C (<0.10) while baboons have a moderate correlation (0.31) that is due entirely to the positive genetic relationship. It is possible that humans, like baboons, could have a genetic correlation between these variables, but the genetic component is not apparent from the phenotypic correlation because the genetic and environmental influences cancel. The genetic correlations of VLDL/LDL and HDL cholesterol with cholesterol turnover rate also were both negative in baboons. These data suggested that the phenotypic relationship between VLDL/LDL and HDL in baboons is the result of a genetic effect on a metabolic process. The gene may affect excretion of bile acids which, in turn, could affect HDL and/or LDL cholesterol turn-over or transfer of cholesterol between lipoprotein fractions. Whatever the mechanism, this study showed that analysis of genetic contributions can help in developing hypotheses about mechanisms that affect cholesterol metabolism and can aid in identifying metabolic relationships when phenotypic correlations are weak.

Determination of genetic effects in baboons have also aided in under-standing the metabolic processes which affect HDL [52]. Significant differences existed among sire progeny groups for HDL cholesterol production rate using a two-pool model of cholesterol metabolism. Strong inverse correlations were found between the means of the sire progeny group HDL cholesterol concentrations and the mass of the rapidly misci-ble pool of cholesterol (pool A) and the rate constants for the movement of cholesterol out of pool A. This outcome suggests that movement of cholesterol out of pool A (excretion from the body) is under genetic con-trol and may be the primary mechanism for controlling HDL cholesterol concentrations. In this case, the difference in HDL cholesterol concentrations among sire progeny groups was due to a genetic effect on catabolism not production, a finding consistent with the suggestion that HDL cholesterol is the preferred precursor for bile acid synthesis [52].

Kushwaha et al. [53] investigated LDL metabolism in high and low responder baboons by studying the turnover in vivo of radioiodinated VLDL and LDL. In response to a diet containing 1.7 mg of cholesterol/kcal, plasma LDL cholesterol and apo B concentrations were higher in high responders than in low responders, 61 and 22 vs. 151 and 74 mg/dl, respectively. An increased proportion of VLDL apo B was converted into LDL apo B and a greater rate of LDL apo B production independent of VLDL were found in the high responders. These results suggested that the LDL concentrations in baboons were related to genetic effects on the pathway(s) for LDL production. It is not clear whether this would be an effect on secretion of apo B-containing particles from the liver or whether it may be an effect on the LDL receptor-mediated clearance of newly secreted VLDL and its conversion products.

In 1984, Kammerer et al. [54] looked at the heritability of TPC, HDL and LDL cholesterol concentrations in the same group of baboons but they compared animals with more extreme responses than those studied by Flow and reached somewhat different conclusions. The relative contribution of heredity and environment were found to change with age with the realized heritability of TPC and the genetic correlation of TPC with LDL and HDL cholesterol concentrations in the low responders being greatest at age 6 months and greatest in the high responders at 2 years of age. Heritabilities of TPC calculated for each age indicated that as the animals mature, genetic factors account for more of the phenotypic vaiation in the high responders but account for less in the low responders. Heritability of TPC, LDL and HDL cholesterol concentrations were similar to that documented in human beings (0.59, 0.36, 0.67, respectively) while the genetic correlation between TPC and HDL cholesterol concentration was higher in the high responders than in the low responders. This outcome may have been a reflection of the relationship between TPC and HDL cholesterol concentration since HDL is a large proportion of TPC in baboons, or it may be due to major genes influencing HDL cholesterol concentrations. Major genes are those individual polymorphic loci that exert large effects on phenotype. In fact, if major genes influence lipoprotein concentrations, then it could be possible to identify those people who are genetically predisposed to hyperlipoproteinemia versus those who are environmentally predisposed.

To determine if genes were affecting lipoprotein cholesterol concentrations in these baboons, MacCluer and co-workers [55–57] evaluated the traits by sibship variance tests and other preliminary methods as well as by complex segregation analysis. The preliminary tests indicated that there is no major gene effect on TPC when either a chow or atherogenic diet is fed and provided weak evidence that there is a major gene affecting apo B. A major gene influences HDL cholesterol concentration in animals fed the chow diet but becomes less apparent in animals fed the atherogenic diet. When the atherogenic diet was fed, a major gene effect was noted for VLDL/LDL which could not be detected on the chow diet [55, 56]. Complex segregation analysis done on the data generated for the chow diet period revealed that a major gene does affect HDL cholesterol concentrations. The gene accounts for 22% of the total phenotypic variation in HDL cholesterol concentration, with a polygenic component accounting for 24% and the environment accounting for 54%. A recessive major gene does influence VLDL/LDL on the atherogenic diet accounting for 11% of the phenotypic variation, but the polygenic model fits best for the values on the atherogenic diet [56, 57].

Summary and Conclusion

Nonhuman primates have become widely accepted as important models for the study of experimental atherosclerosis. Dietary induction of atherosclerosis in nonhuman primates is invariably associated with hypercholesterolemia and hyperlipoproteinemia. The studies covered in this review defined some of the characteristics of plasma lipoproteins in nonhuman primates and describe available information about known genetic components involved in regulation of lipoproteins in these animals. In many cases presented, genetic traits of the plasma lipoprotein regulatory systems become apparent only after the appropriate environmental stimulus, usually dietary cholesterol, is applied. In only two animals, the rhesus monkeys of Morris and Fitch [29], was the genetic trait readily apparent even in the absence of dietary cholesterol. It seems apparent that genes may regulate lipoprotein cholesterol concentrations by regulating some aspect of lipoprotein metabolism per se but they also may regulate the response to environmental stimuli that could indirectly affect lipoprotein metabolism. Although genetic studies in monkeys are in their infancy, the gene for apolipoprotein A-I already has been identified in a number of species as a potential regulatory site related to dietary sensitivity to hypercholesterolemia. The next few years promise to bring much more information about the specific role of this gene and of many others in regulating cholesterol and lipoprotein metabolism.

References

1   Goldberg, A. C.; Schonfeld, G.: Effects of diet on lipoprotein metabolism. Annu. Rev. Nutr. *5:* 195–212 (1985).
2   Mahley, R. W.: Atherogenic hyperlipoproteinemia. The celluar and molecular biology of plasma lipoproteins altered by dietary fat and cholesterol. Med. Clin. North. Am. *66:* 375–402 (1982).
3   Rudel, L. L.; Parks, J. S.; Carroll, R. M.: Effects of polyunsaturated vs. saturated dietary fat on nonhuman primate HDL; in Perkins, Visek, Dietary fats and health, pp. 649–666 (American Oil Chemist's Society, Champaign 1983).
4   McGill, H. C.: Relationship of dietary cholesterol to serum cholesterol concentration and to atherosclerosis in man. Am. J. Clin. Invest. *32:* 2644–2702 (1979).
5   Lipid Research Clinics Program: The Lipid Research Clinics Coronary Primary Prevention Trial Results. I. Reduction in incidence of coronary heart disease. J. Am. Med. Assoc. *251:* 351–364 (1984).
6   Lipid Research Clinics Program: The Lipid Research Clinics Coronary Primary Prevention Trial Results. II. The relationship of reduction in incidence of CHD to cholesterol lowering. J. Am. Med. Assoc. *251:* 365–374 (1984).
7   Report of the Working Group on Arteriosclerosis of the National Heart, Lung and Blood Institute: Environmental and genetic factors in atherosclerosis. Arteriosclerosis. *2:* 52–73 (1981).
8   Rudel, L. L.; Lofland, H. B.: Circulating lipoproteins in nonhuman primates. Prim. Med., vol. 9, pp. 224–266 (Karger, Basel 1976).
9   Rudel, L. L.: Plasma lipoproteins in atherogenesis in nonhuman primates; in Kalter, Use of nonhuman primates in cardiovascular disease, pp. 37–57 (University of Texas Press, Austin 1988).
10 Armstrong, M. L.: Atherosclerosis in rhesus and cynomolgus monkeys. Prim. Med., vol. 9, pp. 16–40 (Karger, Basel 1976).
11 Clarkson, T. B.; Hamm, T. E.; Bullock, B. C., et al.: Atherosclerosis in Old World monkeys. Prim. Med., vol. 9, pp. 66–89 (Karger, Basel 1976).
12 Rudel, L. L.; Bond, M. G.; Bullock, B. C.: LDL heterogeneity and atherosclerosis in nonhuman primates. Ann. NY Acad. Sci. *454:* 248–253 (1985).
13 Armstrong, M. L.; Megan, M. B.; Warner, E. D.: Intimal thickening in normocholesterolemic rhesus monkeys fed low supplements of dietary cholesterol. Circ. Res. *34:* 447–454 (1974).
14 McGill, H. C.; McMahan, M. R.; Kruski, A. W., et al.: Relationship of lipoprotein cholesterol concentrations to experimental atherosclerosis in baboons. Arteriosclerosis *1:* 3–12 (1981).
15 Pitts, L. L.; II, Rudel, L. L.; Bullock, B. C., et al.: Sex differences in the relationship of low density lipoproteins to coronary atherosclerosis in *Macaca fascicularis.* Fed. Proc. *35:* 293 (1976).
16 Hamm, T. E.; Kaplan, J. R.; Clarkson, T. B., et al.: Effects of gender and social behavior on the development of coronary artery atherosclerosis in cynomolgus monkeys. Atherosclerosis *48:* 221–233 (1983).
17 Lofland, H. B.; St Clair, R. W.; MacNintch, J. E., et al.: Atherosclerosis in New World primates. Arch. Pathol. *83:* 211–214 (1967).
18 St. Clair, R. W.; MacNintch, J. E.; Middleton, C. C., et al.: Changes in serum cholesterol levels of squirrel monkeys during importation and acclimation. Lab. Invest. *16:* 828–832 (1967).

19   Clarkson, T. B.; Lofland, H. B.; Bullock, B. C., et al.: Genetic control and plasma cholesterol studies on squirrel monkeys. Arch. Pathol. *92:* 37–45 (1971).

20   Lofland, H. B., Jr., Clarkson, T. B.; St. Clair, R. W., et al.: Studies on the regulation of plasma cholesterol levels in squirrel monkeys of two genotypes. J. Lipid. Res. *13:* 39–47 (1972).

21   Jones, D. C.; Lofland, H. B.; Clarkson, T. B., et al.: Symposium: Nutritional Perspectives and Atherosclerosis. Plasma cholesterol concentrations in squirrel monkeys as influenced by diet and phenotype. J. Food. Sci. *40:* 2–7 (1975).

22   Guertler, L. S.; St. Clair, R. W.: In vitro regulation of cholesterol metabolism by low density lipoproteins in skin fibroblasts from hypo-and hyperresponding squirrel monkeys. Biochim. Biophys. Acta. *487:* 458–471 (1977).

23   Lee, J. A.; Morris, M. D.: Characterization of the serum low-density lipoproteins of normal and two rhesus monkeys with spontaneous hyperbetalipoproteinemia. Biochem. Med. *10:* 245–257 (1974).

24   Rudel, L. L.; Greene, D. G.; Shah, R.: Separation and characterization of plasma lipoproteins of rhesus monkeys *(Macaca mulatta).* J. Lipid. Res. *18:* 734–744 (1977).

25   Nelson, C. A.; Morris, M. D.: A new serum lipoprotein found in many rhesus monkeys. Biochem. Biophys. Res. Commun. *71:* 438–444 (1976).

26   Rudel, L. L.; Shah, R.; Greene, D. G.: Study of the atherogenic dyslipoproteinemia induced by dietary cholesterol in rhesus monkeys *(Macaca mulatta).* J. Lipid. Res. *20:* 55–65 (1979).

27   Clarkson, T. B.; Bond, M. G.; Bullock, B. C., et al.: A study of atherosclerosis regression in *Macaca mulatta.* IV. Changes in coronary arteries from animals with atherosclerosis induced for 19 months and then regressed for 24 or 48 months at plasma cholesterol concentrations of 300 or 200 mg/dl. Exp. Mol. Pathol. *34:* 345–368 (1981).

28   Eggen, D. A.: Cholesterol metabolism in groups of rhesus monkeys with high and low response of serum cholesterol to an atherogenic diet. J. Lipid. Res. *17:* 663–673 (1976).

29   Morris, M. D.; Fitch, C. D.: Spontaneous hyperbetalipoproteinemia in the rhesus monkey. Biochem. Med. *2:* 209–215 (1968).

30   Rudel, L. L.; Pitts, L. L.; II, Nelson, C.: Characterization of plasma low density lipoproteins of nonhuman primates fed dietary cholesterol. J. Lipid. Res. *18:* 211–222 (1977).

31   Fredrickson, D. S.; Goldstein, J. L.; Brown, M. S.: The familial hyperlipoproteinemias; in Stanbury, Wyngaarden, Fredrickson, The metabolic basis of inherited disease, pp. 604–655 (McGraw-Hill, New York 1978).

32   Morris, M. D.; Greer, W. E.: Familial hyperbetalipoproteinemia (type II hyperlipoproteinemia) in the rhesus monkey; in Jones, Atherosclerosis: proceedings of the second international symposium, p. 192 (Springer, New York 1970).

33   Morris, M. D.; Rudel, L. L.; Walls, R. C., et al.: The effects of dietary cholesterol in plasma cholesterol and lipoproteins in progeny of rhesus monkeys with naturally occurring hypercholesterolemia. Circulation *51–52:* suppl. II, p. 15 (1975).

34   Guertler, L. S.; St. Clair, R. W.: Low density lipoprotein receptor activity on skin fibroblasts from rhesus monkeys with diet-induced or spontaneous hypercholesterolemia. J. Biol. Chem. *225:* 92–99 (1980).

35   Brown, M. S.; Goldstein, J. L.: A receptor-mediated pathway for cholesterol homeostasis. Angew. Chem. *25:* 583–602 (1986).

36   Morris, M. D.; Greer, W. E.: Hyperalphalipoproteinemia in cholesterol-fed rhesus monkeys. Fed. Proc. *31:* 727a (1982).

37   Tall, A. R.; Small, D. M.; Atkinson, D., et al.: Studies on the structure of low density

lipoproteins isolated from *Macaca fascicularis* fed an atherogenic diet. J. Clin. Invest. *62:* 1354–1363 (1978).

38 Rudel, L. L.; Pitts, L. L.; II.: Male-female variability in the dietary cholesterol-induced hyperlipoproteinemia of cynomolgus monkeys *(Macaca fascicularis).* J. Lipid. Res. *19:* 992–1003 (1978).

39 Johnson, F. L.; St. Clair, R. W.; Rudel, L. L.: Studies of the production of low density lipoproteins by perfused livers from nonhuman primates: Effect of dietary cholesterol. J. Clin. Invest. *72:* 221–236 (1983).

40 Clarkson, T. B.; Kaplan, J. P.; Adams, M. R.: The role of individual differences in lipoprotein, artery wall, gender and behavorial responses in the development of atherosclerosis. Ann. NY. Acad. Sci. *454:* 28–45 (1985).

41 St. Clair, R. W.; Wood, L. L.; Clarkson, T. B.: Effect of sucrose polyester on plasma lipids and cholesterol absorption in African green monkeys with variable hyper-cholesterolemic response to dietary cholesterol. Metabolism *30:* 176–183 (1981).

42 Rudel, L. L.; Reynolds, J. A.; Bullock, B. C.: Nutritional effects of blood lipid and HDL cholesterol concentrations in two subspecies of African green monkeys *(Cercepithecus aethiops).* J. Lipid. Res. *22:* 278–286 (1981).

43 Sorci-Thomas, M.; Prack, M. M.; Dashti, N., et al.: Apolipoprotein (Apo) A-I production and mRNA abundance explain plasma Apo A-I and high density lipoprotein differences between two nonhuman primate species with high and low susceptibilities to diet-induced hypercholesterolemia. J. Biol. Chem. *263:* 5183–5189 (1988).

44 Parks, J. S.; Lehner, N. D. M.; St Clair, R. W., et al.: Whole-body cholesterol metabolism in cholesterol-fed African green monkeys with a variable hypercholesterolemia response. J. Lab. Clin. Med. *90:* 1021–1034 (1977).

45 De La Pena, A.; Matthijssen, G.; Goldzieher, J. W.: Normal values for blood constituents of the baboon. II. Lab. Anim. Sci. *22:* 249–257 (1972).

46 Kushwaha, R. S.; Hazzard, W. R.; Harker, L. A., et al.: Lipoprotein metabolism in baboons. Effect of feeding cholesterol-rich diet. Atherosclerosis *31:* 65–76 (1978).

47 McGill, H. C., Jr., McMahan, C. A.; Kushwaha, R. S., et al.: Dietary effects on serum lipoproteins of dyslipoproteinemic baboons with high HDL₁. Arteriosclerosis *6:* 651–663 (1986).

48 Babiak, J.; Gong, E. L.; Nichols, A. V., et al.: Characterization of HDL and lipoproteins intermediate to LDL and HDL in the serum of pedigreed baboons fed an atherogenic diet. Atherosclerosis *52:* 27–45 (1984).

49 Flow, B. L.; Cartwright, T. C.; Kuehl, T. J., et al.: Genetic effects on serum cholesterol concentrations on baboons. J. Hered. *72:* 97–103 (1981).

50 Flow, B. L.; Mott, G. E.: Genetic mediation of cholesterol metabolism in the baboon *(Papio cynocephalus).* Atherosclerosis *41:* 403–414 (1982).

51 Flow, B. L.; Mott, G. E.; Kelley, J. L.: Genetic mediation of lipoprotein cholesterol and apoprotein concentrations in the baboon (*Papio* sp.). Atherosclerosis *43:* 83–94 (1982).

52 Flow, B. L.; Mott, G. E.: Relationship of high density lipoprotein cholesterol to cholesterol metabolism in the baboon (*Papio* sp.). J. Lipid. Res. *25:* 469–473 (1984).

53 Kushwaha, R. S.; Barnwell, G. M.; Carey, K. D., et al.: Metabolism of apoprotein B in selectively bred baboons with low and high levels of low density lipoproteins. J. Lipid. Res. *27:* 497–507 (1986).

54 Kammerer, C. M.; Mott, G. E.; Carey, K. D., et al.: Effects of selection for serum cholesterol concentrations on serum lipid concentrations and body weight in baboons. Am. J. Med. Genet. *19:* 333–345 (1984).

55   MacCluer, J. W.; Kammerer, C. M.: Complex segregation analyses of HDL-C in baboons (abstract). 7th Int. Congr. on Human Genetics, Berlin. 7: 732 (1986).
56   MacCluer, J. W.; Kammerer, C. M.; VandeBerg, H. L., et al.: Detecting genetic effects on lipoprotein phenotypes in baboons: a review of methods and preliminary findings. Genetica (in press).
57   Kammerer, C.; MacCluer, J.; Mott, G.: Evidence for major genes influencing serum lipoprotein and apolipoprotein concentrations in baboons: results of SEDA and PAP analyses. Am. J. Hum. Genet. 39: 709 (1986).

Kathy Laber-Laird, DVM, Arteriosclerosis Research Center, Departments of Comparative Medicine and Biochemistry, The Bowman Gray School of Medicine of Wake Forest University, Winston-Salem, NC 27103 (USA)

Lusis A J, Sparkes S R (eds): Genetic Factors in Atherosclerosis: Approaches and
Model Systems. Monogr Hum Genet. Basel, Karger, 1989, vol 12, pp 189–222

Atherosclerosis in the Mouse[1]

Brian Y. Ishida, Beverly Paigen[2]

Children's Hospital Oakland Research Institute, Children's Hospital Medical Center,
Oakland, Calif., USA

Introduction

It has long been desirable to study atherosclerosis in the mouse because genetic mutants with altered physiological responses have proved
so useful in analyzing the underlying mechanisms in many biological
systems. This species has the most sophisticated genetic system known
among experimental mammals and its use would provide a system for
examining the genetic, biochemical, and molecular mechanisms of atherogenesis. Only in recent years, however, have appropriate experimental
regimens been developed that reliably result in reproducible atherosclerotic lesions in the mouse. This review will summarize the characteristics of these regimens, the genetic tools available in the mouse, the
genes in lipid transport and atherosclerosis susceptibility that have so far
been identified, and the lipoprotein profile of the mouse. Many of the
genes identified have parallels in humans and may provide experimental insights into human genetic variation affecting atherosclerosis
susceptibility.

[1] Supported in part by grant HL-32087 from the Heart, Lung, and Blood Institute and
by Biomedical Research Support Grant RR05467 from the National Institutes of Health, by
grant 86–1387 from the American Heart Association with funds contributed in part by the
Alameda, Orange, and Santa Barbara County Chapters, and by grant 87–N119 from the
California Affiliate of the American Heart Association.

[2] The authors wish to thank Renee LeBoeuf for her thorough and helpful review of the
manuscript.

Historical Development of the Mouse Model for Atherosclerosis Research

Early attempts to use the mouse as an experimental model for atherosclerosis research were of limited success. Reasons for this, as summarized in a review of animal models [38], were: (a) inconsistent and variable results with respect to the size, location and progression of lesions; (b) high mortality of mice on diets used to promote lesions; (c) a lesion pathology apparently unlike that found in humans; (d) small animal size which presented technical problems, and (e) a lipoprotein profile that appeared to differ significantly from humans [7]. Fortunately, through the development of improved diets, use of appropriate mouse strains, and better characterization of lesion pathology and lipoproteins, the mouse has become established as a useful tool in characterizing the genetic aspects of heart disease.

Reproducibility of Lesion Formation

Initially, investigators reported that the formation of atherosclerotic lesions in mice was highly variable and inconsistent within the same treatment group [2, 15, 16, 49, 88, 89, 94]. However, Roberts and Thompson [78, 79, 86] reasoned that the variability observed might be genetic in origin since randomly bred mice had been used. These investigators hypothesized that the formation of lesions would be reproducible within an inbred strain. Furthermore, they reasoned that strains should differ in their relative susceptibility to atherosclerosis. Based on these assumptions, Roberts and Thompson [78, 79, 86] surveyed several inbred strains of mice: they found that lesion formation was reproducible within a strain and that inbred strains do differ in atherosclerosis susceptibility. The use of inbred strains, with their advantages of being homozygous at each gene and genetic identity between members of the same strain, contributed a major step forward in the development of the mouse as a model for atherosclerosis research.

Development of Nonlethal Diets

Investigators initially produced lesions in mice using diets containing 30% by weight of saturated fat and 5% by weight of cholesterol. Although mice were gradually introduced to this diet by increasing the dietary fat during several weeks, many animals died [16, 88, 89]. Some investigators observed a 'cage effect' in mortality and reported that death was usually due to respiratory infections [57, 66]. One explanation of the high mortality caused by these diets was that a high-fat diet suppressed the immune response in mice [13, 20, 34, 39, 47, 48, 54, 55]. This difficulty

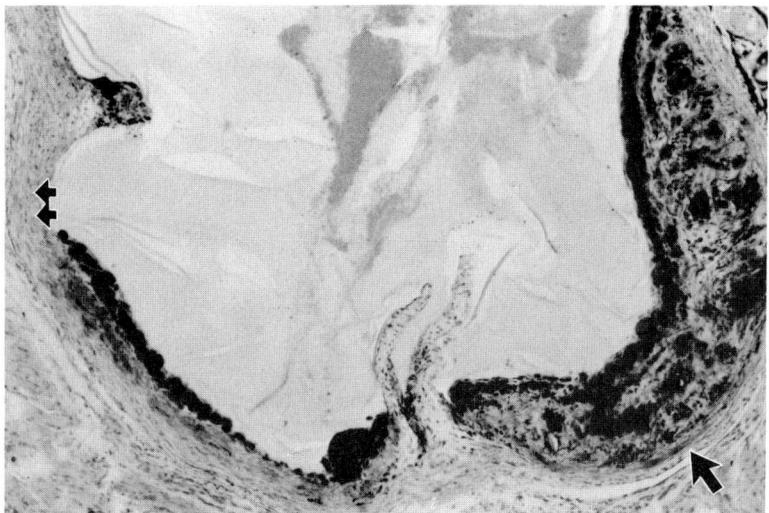

Fig. 1. Advanced atheromatous lesion in strain C57BL/6. The lesion on the left is an advanced fatty streak. The lesion on the right is an advanced atheromatous plaque with cellular debris and cholesterol crystal deposits in the deeper layers of the lesion. The arterial wall underneath the lesion has thinned (single arrow) compared to normal arterial wall (double arrow) and is deficient in elastin fibers. The mouse was fed atherogenic diet for 6 months. Oil red O, 135 × .

was resolved by the discovery that lesion formation could be induced in genetically susceptible mice at levels of dietary fat that do not cause increased susceptibility to infection [54, 55, 66]. We have found such a diet containing 15% fat (by weight), 1.25% cholesterol, and 0.5% cholic acid with a polyunsaturated/saturated ratio of 0.7 [66, 69].

Similarity of Murine and Human Lesion Pathology

A concern of investigators using the mouse as a model for atherosclerosis research was the apparent lack of similarity of murine lesions to those seen in humans. Lesions of mice fed diets containing 30% by weight saturated fat and 5% by weight cholesterol, appeared as fatty streaks rather than mature complicated plaques with fibrous caps characteristic of human atheromas. We now understand that this difference reflected the fact that mice did not survive for sufficiently long periods of time to produce mature lesions. Atherosclerosis-susceptible mice, maintained on a 15% by weight fat and 1.25% cholesterol diet for 6–9 months, form complex atheromatous plaques throughout the arterial tree, histologically similar to those observed in man [87]. Figure 1 depicts a mouse fed our

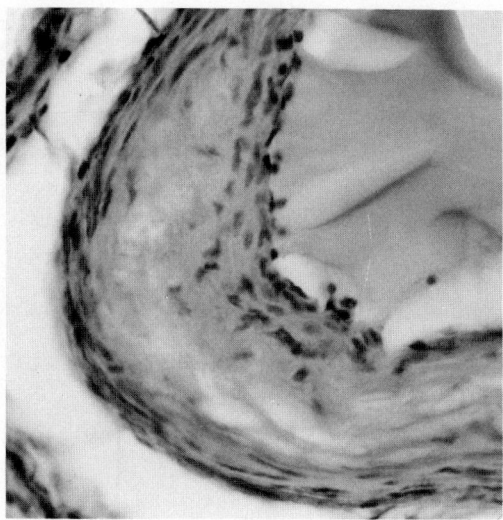

Fig. 2. Atheromatous plaque. This lesion, stained with HE, indicates the cellular nature of the fibrous plaque and the deficiency of cells within the atheroma. The deeper layers of the lesion consist mainly of cellular debris and lipid accumulation. C57BL/6 mouse fed diet for 6 months. HE, 550 ×.

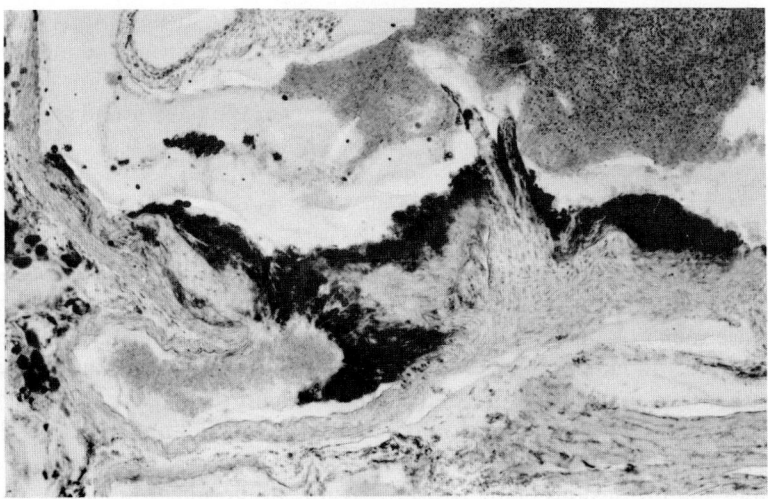

Fig. 3. Occluded coronary artery. The egress of the coronary artery from the lumen of the aorta is shown in the lower center of the photo. The egress is blocked by a thick atherosclerotic lesion. C57BL/6 mouse fed diet for 9 months. Oil red O, 135 ×.

atherogenic diet for 6 months with an advanced fatty streak and a large, atheromatous lesion in the aorta. A thick cap covers the atheroma, which contains elastin fiber and cellular debris and cholesterol crystals. Beneath the lesion, the arterial wall has thinned considerably and contains fewer elastin fibers than normal. Staining with hematoxylin and eosin shows that the cap is cellular and fibrous (fig. 2) and a reduced number of cells reside within the deeper layers of the lesion. Staining with Verhoeff's stain indicates that the fibrous cap contains considerable collagen. Mice from the atherosclerosis-susceptible strain C57BL/6 die after 30–40 weeks' consumption of our high-fat diet, and autopsy shows aortic aneurysms and occluded coronary arteries (fig. 3). Their pathological findings are similar to those which occur in humans and establishes the mouse as a relevant animal model for atherosclerotic disease.

Animal Size

The size of the mouse is both a problem and an advantage. It is a problem because conventional assays for lipoproteins and lipids, developed for humans and other larger species, must be scaled down by factors of 10–100 and because measurements of atherosclerotic lesions must be done microscopically. With the advent of more sensitive methods for protein (silver staining and immunoblotting) and lipid (HPTLC) analysis and the development of centrifuges and rotors with small volume capacities, it is possible to obtain a total lipoprotein profile from 200 μl plasma available from the tail bleeding of a single live mouse. The small size of the mouse provides several advantages including low cost, availability, ease of maintenance in large numbers, and reproductive cycle short enough to make genetic crosses feasible.

Lipoprotein Profiles in Human and Mouse

The customary diet of humans contains 30–45% of calories as fat with considerable exogenous cholesterol, while the laboratory chow fed to mice contains 8% of calories as fat and very little cholesterol. Under such circumstances, humans and mice differ in the relative quantities of lipoprotein subclasses, since mice have primarily HDL and humans have primarily LDL and VLDL (table I) [7]. The difference in lipoprotein profile diminishes when mice are fed diets more akin to those of men. The lipoprotein profiles of mice fed a diet containing 30% of calories as fat and 1.25% cholesterol approach that of humans (table I). Furthermore, the quantitative levels of HDL in the two species have similar consequences for atherosclerosis. For example, humans with HDL-cholesterol levels above 65 mg/dl have a greatly reduced risk of coronary heart disease; similarly C3H, a strain of mice that has levels of HDL-cholesterol

Table I. Comparison of lipoprotein-cholesterol in mouse and human (values are given in mg cholesterol/dl plasma)

	Total cholesterol	LDL and VLDL cholesterol	HDL cholesterol
Average human	220	175	45
C57BL/6 mice[a]			
Chow	66 ± 14[b]	4[c]	62 ± 9[b]
Atherogenic diet	192 ± 26	161	31 ± 6
C3H mice[a]			
Chow	84 ± 3	15	69 ± 13
Atherogenic diet	207 ± 32	138	69 ± 9

[a] C57BL/6 mice are susceptible to atherosclerosis; C3H mice are resistant.
[b] Mouse data from Paigen et al. [68]. Values are given as mean ± SEM.
[c] Mouse LDL and VLDL values were obtained by subtracting HDL-cholesterol from total cholesterol so there are no standard errors.

above 65 mg/dl while consuming an atherogenic diet, is also resistant to atherosclerosis [68]. Likewise, humans with HDL-cholesterol levels below 30 mg/dl have an increased risk of atherosclerosis; strain C57BL/6 has HDL-cholesterol levels in the 30–35 mg/dl range and is susceptible to atherosclerosis.

Characteristics of the Mouse Model

Diet

The three laboratories that have evaluated lesion formation in mice used variations of the Thomas-Hartroft diet, first developed to produce atherosclerosis in rats. This diet, which contains by weight 30% cocoa butter, 5% cholesterol and 2% cholic acid, represents a much greater atherogenic insult than the diet consumed by humans. Mice generally refuse this diet, but Roberts and Thompson [78, 86] were able to use it by first offering the diet to weanling mice as a mixture of 7 parts diet to 3 parts laboratory chow for a few days, then as a mixture of 8 parts diet to 2 parts chow, and finally as 9 parts diet to 1 part chow for the remainder of the experiment. Using such a diet, Roberts and Thompson found lesions in the aortic sinus after mice consumed the diet for 6 weeks. However, other investigators using this diet found excessive morbidity and mortality [57, 68]. Paigen et al. [68] tested several mixtures of the diet and breeder chow and found that very little morbidity or mortality occurred within the first 3 months when mice were fed a mixture of 1 part diet to 3 parts breeder chow. This mixture results in a fat content of 15% and cholesterol of 1.25% and is comparable to the fat content of the diet consumed by humans. On this diet, susceptible mice develop fatty streaks as early as 6–7 weeks and more extensive lesions by 10–14 weeks. This diet not only produces atherosclerotic lesions, but it also causes the formation of gallstones and fatty livers in susceptible mice [87].

Although we continue to use this diet [51, 65–72], it is not ideal due to the varying composition of natural ingredients such as oats and corn in commercial breeder chow. Efforts to develop a chemically defined synthetic diet that produces lesions in a convenient length of time are in progress. A diet based on hydrogenated coconut oil and glucose did not produce large enough lesions in the susceptible strain at 3 months, but recently a diet based on cocoa butter and sucrose did produce a lipoprotein profile that is associated with lesion formation [Nishina et al., unpubl.].

Quantitation of Lesions

Roberts and Thompson [78, 79, 86] were the first to compare lesion formation among inbred strains. They used cross sections from the aortic sinus and developed a semiquantitative comparison of lesions based on a 1–3 scoring system that took into account the relative fraction of the aortic perimeter involved in lesion and the number of foam cells. They used cross sections selected randomly from the aortic sinus and evaluated each lesion only once in the cross section where the lesion had the largest size. This system, while able to evaluate a number of inbred strains and rank them in order of atherosclerosis susceptibility, was not precise enough to accurately quantitate lesions in individual mice as required in genetic crosses between resistant and susceptible strains.

Paigen et al. [70] modified this system by defining the region of the aorta to be evaluated more carefully, by selecting cross sections at precisely defined intervals rather than randomly, and by evaluating lesion size using a calibrated microscope eyepiece so that the cross-sectional area in μm^2 could be obtained. The region selected for scoring is 300–350 μm long in a mouse of 22–23 g. The beginning and the end of the region can be recognized by key anatomic features in the cross section [70]. Five sections at equal distance from each other are selected and the number and the size of lesion in each cross section determined. Each animal contributes five separate values to an experimental group. Statistical differences (for example, between strains) can be evaluated in groups as small as 5 mice (25 values). Such group sizes are often used when the strains differ considerably in susceptibility, but larger groups are used when evaluating mice which may be intermediate in phenotype, such as an F_1 generation.

Table II. Atherosclerosis susceptibility among inbred strains of mice

Strain	Method 1[a] – number of lesions ± SEM		Method 2[b] – lesion size ± SEM, μm^2
	week 10	week 14	week 14
C57BR	–	–	7,940 ± 1,330
C57BL/6	1.2 ± 0.3	1.8 ± 0.3	4,200 ± 860
C57L	0.3 ± 0.2	0.8 ± 0.2	6,130 ± 2,700
SM	–	–	4,490 ± 670
SWR	0	–	1,690 ± 280
DBA/2	0	1.0 ± 0.3	200 ± 80
AKR	0	0.7 ± 0.3	44 ± 24
129	0	0.4 ± 0.4	350 ± 120
NZB	0	0	–
BALB/c	0	0	20 ± 15
HRS	–	–	0
A	0	0	0
C3H	0	0	0
SJL	–	–	0
CBA	–	–	0
Peru	–	–	0

[a] Method 1 scored the number of lesions/mouse in a defined area counting each lesion once whether it appeared in one or several aortic cross sections.
[b] Method 2 measured total lesion size in 5 predetermined cross sections from each mouse. Both methods are described in Paigen et al. [70]. Some data have been published [67, 69].

Although the two laboratories used somewhat different methods to quantitate lesions, the agreement between them as to relative susceptibility of strains to atherosclerosis was excellent. Roberts found the order of inbred strains from most to least susceptible to be: C57BR/cdJ, C57L/J, SWR/J, C57BL/6J, C57BL/10J, DBA/1J, A/J, BALB/cJ, C3H/HeJ, and CBA/J. Paigen and co-workers found the order to be C57BL/6J, C57L/J, DBA/2J, AKR/J, 129/J, NZB/J, BALB/cJ, A/J, and C3H/HeJ. The C57 strains are related to each other as are the DBA/1 and DBA/2 strains and also the CBA and C3H strains. (The strain survey by Dr. Roberts appears in his PhD thesis and is quoted by Paigen et al. [69].)

Atherosclerosis Susceptibility among Inbred Strains
The inbred strains that have been examined for susceptibility to atherosclerosis by Paigen and co-workers are shown in table II. The first method of measuring susceptibility used a modification of the Roberts

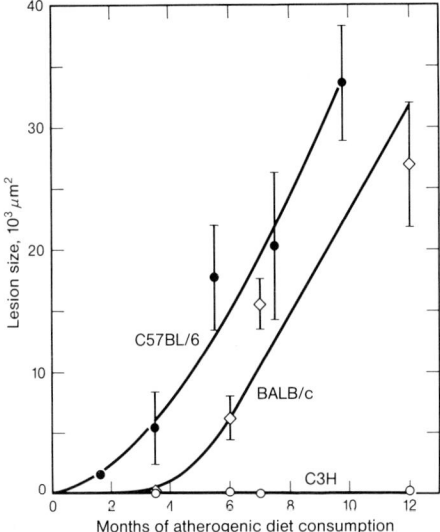

Fig. 4. Comparison of lesion formation strains C57BL/6, BALB/c and C3H. No C57BL/6 mice survived beyond 10 months. BALB/c and C3H mice survived the entire year.

and Thompson scoring method and relies on the number of lesions, counting each lesion singly regardless of its size [69]. The second improved method depends on lesion size as described above [70]. Several strains were examined using both methods.

Originally, these strains were designated as resistant or susceptible to atherosclerosis based on evaluation after 10 or 14 weeks consumption of atherogenic diet. However, each strain has a time course of lesion formation, and susceptibility to atherosclerosis is a relative term that must be considered at a particular time point. For example, C57BL/6 mice are susceptible at 14 weeks, but C3H and BALB/c are resistant. However, if the experiment is carried out for a year instead of 14 weeks, a different picture is obtained (fig. 4). The lesions in C57BL/6 grow and can occlude the coronary arteries. Between 28 and 40 weeks, the mice die displaying major coronary artery disease involvement at postmortem. BALB/c and C3H, the two resistant strains, are similar at 14 weeks, but lesion formation begins to differ at later time points. By 1 year, C3H still has almost no lesions and can be said to be truly resistant (fig. 5). However, BALB/c begins to develop lesions, and by 1 year these involve most of the perimeter of the aorta. The lesions in BALB/c do differ from C57BL/6 in that

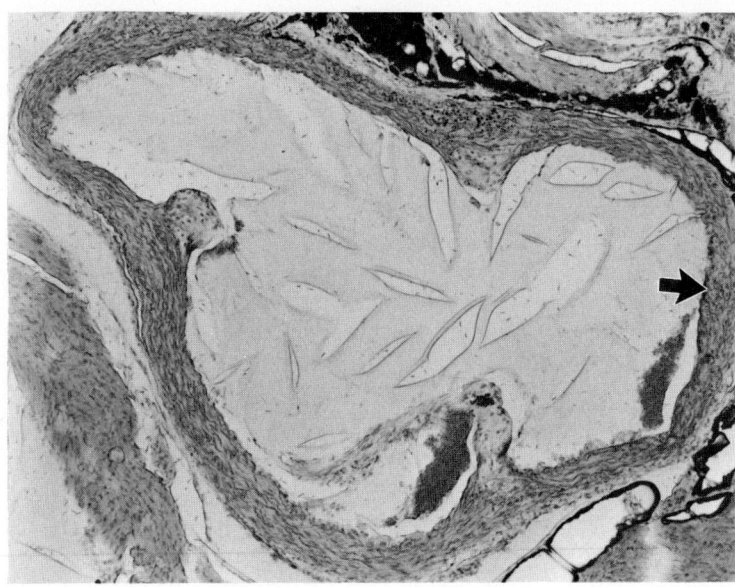

Fig. 5. Aorta of C3H mouse. C3H mice do not develop lesions or develop only tiny pinpoint lesions (see arrow). This particular mouse was fed diet for 7 months and the photo was selected to illustrate the tiny lesion. The C3H mice sacrificed at 12 months were similar to the above photo but lacked any lesion at all. C3H mouse fed diet for 7 months. Oil red O, 135 × .

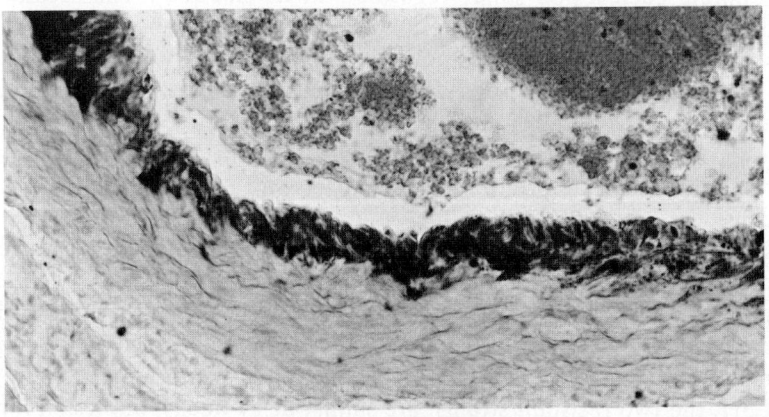

Fig. 6. Lesion in BALB/c mouse. The lesion in this BALB/c mouse, although thick, does not contain much atheromatous material. Furthermore, the artery wall underneath has not thinned nor does the lesion protrude much into the lumen. Thus, the lesion pathology in BALB/c is quite different from C57BL/6 (see fig. 1, 2).

they do not affect the coronary arteries, they do not protrude significantly into the lumen of the aorta and usually they do not have atheromas (fig. 6). Thus, they are unlikely to cause mortality due to coronary artery disease. These data suggest that it is more appropriate to consider BALB/c as intermediate between C57BL/6 and C3H, even though both strains carry the resistant alleles of *Ath-1* as mentioned below. Likewise, some of the strains that look susceptible at 14 weeks can be distinguished from each other if an earlier time point is chosen.

Genetic Variants in Lipid Transport and Atherosclerosis Susceptibility

Finding New Genetic Variants

There are two approaches to find mouse strains which express variations in the lipid transport system. The first is to select one component of that system, such as an apoprotein or lipid-metabolizing enzyme, and search for variants by surveying inbred strains [methodology reviewed in 50]. Such a search will find quantitative variants in a range consistent with survival, similar to the range observed in humans, which are the based for differing susceptibility to common diseases.

The second approach involves examination of the large number of mutant strains with phenotypes which suggest some alteration in lipid transport [26, 50]. With such a mutant, it is possible to examine the consequences when a single component of the lipid transport system is altered without the problems of environment and heterogeneous genetic background that complicate human studies. These mutants often represent an extreme pertubation of lipid transport, and an understanding of their biochemical and molecular mechanisms may contribute greatly to understanding this system.

Variants in Apolipoproteins and Lipid-Metabolizing Enzymes

Many genetic variants in apolipoproteins and lipid-metabolizing enzymes have been found. Their identification, map positions in the mouse, and the homologous map positions in the human are reviewed in another chapter in this volume [53] and will be briefly mentioned here. The structural genes for apo A-I and apo A-II were located by linkage analysis of electrophoretic apoprotein polymorphisms from a survey of 40 inbred mouse strains [52]. The structural genes for apo B and apo E were found by DNA RFLP polymorphisms [51, 77]. A variant for quantitative levels of apo A-IV mRNA in response to an atherogenic diet was reported [95]. Variants have not been identified for apo C.

Quantitative differences in the levels of apo A-II, apo B-100, apo B-48, and apo E have been reported [18, 51]. A quantitative variation in apo A-II has been identified among mouse strains, which affects its rate of synthesis and appears to be controlled by a genetic element linked to the apo A-II structural gene on chromosome 1. Quantitative differences in the levels of LDL and VLDL [51] and HDL [68] have been reported. HDL differences were linked to susceptibility to atherosclerosis in some inbred strains [68]. Variants for lipoprotein lipase, hepatic lipase, LCAT, cholesteryl ester transfer protein, HMG-CoA synthase, and HMG-CoA reductase, have also been found and mapped [4, 50, 53, Lusis and LeBoeuf, pers. commun]. No variants have yet been reported for the LDL or HDL receptor proteins.

Genetic Variants Affecting Atherosclerosis Susceptibility

Ath-1

The first gene described that alters susceptibility to atherosclerosis is *Ath-1* with the alleles *r* for resistant and *s* for susceptible [65, 68, 71]. This gene determines the level of HDL-cholesterol in mice consuming an atherogenic diet as well as the susceptibility to lesion formation. The susceptible and resistant strains do not differ in total cholesterol or in LDL and VLDL lipoproteins when consuming an atherogenic diet, but they do differ in HDL-cholesterol. Strain C57BL/6 carries the susceptible allele and strains BALB/c and C3H carry the resistant allele. Phenotypes of atherosclerosis susceptibility and HDL-cholesterol levels are determined by the same genetic factor (or by closely linked genetic factors at a maximum distance of 1.7 cM from each other), as demonstrated using recombinant inbred strains derived from atherosclerosis-susceptible and resistant progenitor strains. In 47 out of 47 recombinant inbred strains, atherosclerosis-susceptible strains exhibited decreased HDL levels; in contrast, resistant strains maintained normal HDL levels. This concordance of phenotypes indicates that characteristics of resistance and HDL levels are determined by the same or by closely linked genes. Recombinant inbred strains and backcrosses were used to demonstrate that the atherosclerosis difference was due to a single gene. This was also established among the 47 RI strains. If a phenotype is determined by a single gene, then each recombinant inbred strain will resemble one progenitor or the other, and the distribution of phenotypes among the recombinant inbreds will be about 1:1. In contrast, if a phenotype is determined by two or more genes, then recombinant inbred strains will exhibit new or intermediate phenotypes and the distribution will not be 1:1.

The data from the recombinant inbred strains derived from BALB/c and C57BL/6 (CXB) or from C57BL/6 and C3H (BXH) suggested that *Ath-1* had a map position near the gene for apo A-II. Conventional crosses, carried out to confirm this map position, demonstrated that *Ath-1* was $4.9 \pm$ (SE) 1.8 cM away from the gene determining apo A-II on chromosome 1 [65, 71]. This map distance was based on 7 crossovers among 144 tested chromosomes.

While consuming the atherogenic diet, the HDL-cholesterol levels in *Ath-1*$^{s/s}$ mice are 30–35 mg/dl, similar to the 30 mg/dl considered as low HDL-cholesterol for humans (hypoalphalipoproteinemia). The HDL-cholesterol levels in *Ath-1*$^{r/r}$ mice were 65–70 mg/dl, close to the 70 mg/dl considered to be high HDL-cholesterol for humans (hyperalphalipo-proteinemia). The F_1 progeny derived from a cross between C57BL/6 and C3H were intermediate in HDL-cholesterol and susceptibility to athero-sclerosis. Both hypoalphalipoproteinemia and hyperalphalipoproteine-mia are traits known to segregate as single genes in human families [25, 80]. Thus, the linkage of the *Ath-1* gene in mouse suggests a possible map location near apo A-II for consideration in these families.

The studies for *Ath-1* primarily used female mice because there were sex differences. Male C57BL/6 mice had HDL-cholesterol levels that were higher than females and male mice were less susceptible to athero-sclerosis. In fact, comparison of HDL-cholesterol and lesion formation in three strains of mice using males, females, and testosterone-treated females, which resembled males in HDL levels, showed a high degree of correlation between HDL-cholesterol and lesion formation ($r = -0.93$) [67]. However, another laboratory using C57BL/6 males showed that males do exhibit low HDL-cholesterol levels if the atherogenic diet is more severe than the one used in our laboratory (5% cholesterol rather than 1.25%) [42].

Ath-2

Strain A, which is closely related to strain C3H, is also resistant to atherosclerosis, and as in C3H, female mice from strain A have high levels of HDL-cholesterol when consuming an atherogenic diet. Thus, the large set of 45 RI strains derived from C57BL/6 and A was examined to determine if these two parental strains also differed at *Ath-1*. The data suggested that these strains differed at *Ath-1* as expected and also differed at a second gene determining atherosclerosis susceptibility, which was named *Ath-2* [73]. *Ath-2* has the same phenotype as *Ath-1* in that strains

Table III. Interaction of *Ath-1* and *Ath-2* to produce atherosclerosis susceptibility phenotypes

Ath-1 alleles	Ath-2 alleles	Strains	Atherosclerosis phenotype
rr	rr	A	resistant
rr	ss	C3H, BALB/c	resistant
ss	rr	10 A × B RI strains[a]	resistant
ss	ss	C57BL/6, 13 A × B strains[a]	susceptible
sr	ss	F₁ (C57BL/6 × C3H)	intermediate
ss	sr	F₁ (C57BL/6 × A × B-5)	resistant

[a] These strains are listed in Paigen et al. [73].

carrying the resistant allele do not get lesions in the aorta at 14 weeks and have high levels of HDL-cholesterol, while strains carrying the susceptible allele have low HDL-cholesterol and are susceptible to lesions. The map position of *Ath-2* is not known, but mapping studies have eliminated linkage to apo A-I, apo A-II, apo B, and apo E as well as many other genetic markers that have been typed in the A × B recombinant inbred set.

Ath-1 and *Ath-2* interact with each other so that having resistant alleles at either locus results in a strain that is resistant to atherosclerosis (table III). In order for a strain to be susceptible, it must carry susceptible alleles at both loci. The interactions of this two gene system is an example of what may occur in human type III hyperlipoproteinemia, which involves the effect of an apo E variant on coronary heart disease risk. Three electrophoretic variants of the apo E structural gene are known and these are designated E2, E3, and E4. The E2 isoform has a reduced binding affinity for the LDL receptor. About 90% of individuals with type III hyperlipoproteinemia are found to be homozygous for E2/E2. However, homozygosity for E2/E2 is not sufficient by itself to cause elevated levels of lipoprotein and increased heart disease risk because most subjects with E2/E2 genotype have normal lipoprotein levels [14]. This puzzling observation could be explained by a two-gene system such as *Ath-1* and *Ath-2*. In order to have elevated lipoprotein levels, an individual would have to be homozygous E2/E2 and also carry susceptible alleles at a second, as yet unknown, locus.

In similar fashion, an RFLP has been found near or within the apo A-I/apo C-III complex. Persons with the less common allele at this site are more likely to have low levels of HDL-cholesterol and increased risk of heart disease. However, some persons with this allele have normal levels of HDL-cholesterol, probably due to a second gene which also affects HDL levels [64].

Strains C57BR and CBA

The original pair of atherosclerosis-susceptible and resistant strains, chosen for analysis by Roberts and Thompson [78, 79], was C57BR and CBA, strains that are closely related to C57BL/6 and C3H, respectively. Several biochemical differences between C57BR and CBA that might account for the difference in atherosclerosis susceptibility have been found: plasma cholesterol levels [78, 79, 90], ratio of apo E/total lipoprotein [57], level of beta- and alpha-migrating lipoproteins [5], and phosphatidylethanolamine to free cholesterol ratio of VLDL, IDL, and LDL [40]. However, it is not clear which of these differences are related to the difference in atherosclerosis susceptibility or even if the susceptibility difference is due to a single gene. Roberts and Thompson [79] carried out a conventional cross and showed that the elevated cholesterol levels were due to two or more genes.

Although several laboratories have examined these strains in detail and made significant contributions to our understanding of mouse lipoproteins, a major gene has not been identified [5, 40, 59, 78, 79, 90]. There are several reasons why the analysis of this pair of strains was less fruitful than an analysis of the C57BL/6 and C3H pair, the C57BL/6 and BALB/c pair or the C57BL/6 and A pair. The reasons are worth examining in some detail because they illustrate how to best use the genetic resources of the mouse model. First, C57BR is a poor breeder which made obtaining sufficient numbers of mice difficult and caused delays in experiments. Breeding performance of common inbred strains is available in the Jackson Laboratory Handbook (see List of Resources below). The price of a mouse provides some indication of breeding performance since the poor breeders are more expensive. Since the set of related strains C57BR, C57L, C57BL/6 and C58 are all susceptible to atherosclerosis, a better breeder could have been chosen from among this set as the atherosclerosis-susceptible strain. A second and more important reason, however, is that a set of recombinant inbred strains do not exist between C57BR and CBA. Had there been a recombinant inbred set, the investigators would have been able to quickly determine (a) whether the difference in atherosclerosis susceptibility was due to a single gene, (b) which of the many biochemical differences between the two strains cosegregated with atherosclerosis susceptibility, and (c) if a single gene was involved, and the tentative map position.

The Lymphoproliferative Mutant in Strain MLR

A mutation causing an autoimmune disease (*lpr* for lymphoproliferative) arose in strain MLR. MLR/1 mice develop massive proliferation of
T lymphocytes and die between 5 and 7 months of age. If MLR/1 and its
normal congenic strain MLR are placed on a diet containing 1%
cholesterol, 0.3% cholic acid, and 20% fat, mice carrying *lpr* developed
atheromatous lesions in the aorta and renal arteries while the normal
mice do not. These data suggest that the abnormal immune response in
lpr mice affects the atherosclerotic process [54, 55].

Mutants with Defective Platelet Function

Several mutants have been characterized with prolonged bleeding
times and defective platelet granules. These mutants – *beige*, *light ear*,
pale ear, *maroon*, and *ruby eye* – have cell organelle defects in lysosomes,
melanosomes and platelet granules. The mutants exist as congenic or
coisogenic strains to C57BL/6. Comparison of atherosclerosis susceptibility in these mutants and the C57BL/6 strain shows that *pale ear* develops
atherosclerosis to the same degree as do C57BL/6; thus, normal platelet
function is not essential to the formation of atherosclerotic lesions.
However, the remaining four mutants had a reduced susceptibility to lesion formation; thus, some specific component of platelet function does
have an effect on lesion formation. The level of serotonin, found in the
dense granules, was not correlated to atherosclerosis, but the level of
platelet thrombospondin, found in the alpha granules, showed an inverse
correlation with atherosclerosis ($r = -0.88$) [72]. *Pale ear* thrombospondin levels and atherosclerosis comparable to C57BL/6, and *ruby eye* and
light ear had the highest levels of thrombospondin and were resistant to
atherosclerosis.

Mutants with Altered Lipid Transport

Several hundred mutants that arose in an inbred strain and have been
saved as congenic or coisogenic strains exist. Those known to affect some
aspect of lipid metabolism are briefly described; a more complete description is available in Green [26].

Obesity and diabetes: The most common type of mutants with altered
lipid metabolism are those with obesity or diabetes [reviewed in 30]. The
obese mutations include *fat*, *obese*, *tubby*, and two alleles at the agouti
locus called yellow (A^y) and viable yellow (A^{vy}) [26]. Some of these mutations are obese due to an increased concentration of lipid in each fat cell

and others'due to an increased number of fat cells [37]. In addition, the mutant *diabetes* is obese and is a better model of human obesity than of human diabetes. A better model of human insulin-dependent diabetes is the strain NOD (nonobese diabetic) [46]. In NOD mice, diabetes is determined by at least three genes, one which maps near the major histocompatability locus on chromosome 17 and one which maps near the gene determining apolipoprotein A-I structure on chromosome 9 [76]. The *diabetes* and *obese* mutants and the NOD strain have been studied extensively [9–11, 30, 37, 46, 76] and will not be reviewed in detail here.

A (yellow and viable yellow): The agouti locus (a), which maps on chromosome 2, affects coat color by altering the distribution and amount of yellow and black pigment. Among the many known alleles, the two that cause yellow coat color also cause obesity . The increase is in fat cell size not in fat cell number [37]. For the allele yellow (A^y), homozygotes die *in utero*. Heterozygotes become obese and sterile after the first few months. The increased fat is due to fat cell hypertrophy and these mice arc obcse even if the diet is restricted so that the body weight is normal [30]. The allele *viable yellow* (A^{vy}) resembles yellow except that homozygotes do not die. Both homozygotes and heterozygotes are obese [30]. Both yellow and viable yellow are more susceptible to tumors and less likely to reject grafts.

Ad (adult obesity and diabetes): This mutant was found in wild-derived mice and maps to chromosome 7 [93].

ald (adrenocortical lipid depletion): This mutant arose in the AKR strain and maps to chromosome 1. At puberty, the adrenal cortex is depleted of cholesterol esters, but not of free cholesterol or triglycerides [17]. The lipid depletion requires intact gonads and hypophysis. The site of action is not the adrenal since adrenals from *ald* mice transplanted into normal mice do not undergo lipid depletion [17, 85].

db (diabetes): This mutant arose in the C57BL/6Ks strain and maps to chromosome 4. Homozygotes are sterile so the mutant is maintained in a heterozygous state. The expression of diabetes depends on its genetic background; when transferred into another strain it is expressed as obesity [9, 12]. Apparently the mutation causes insulin resistance. When it is present in strain C57BL/6Ks other genes interact with *diabetes* in such a way as to cause severe diabetes; however, when *diabetes* is transferred by repeated backcrosses with other strains such as C57BL/6J and 129, the genetic background interacts with *diabetes* to cause obesity. The obesity is present even when food intake is restricted to normal food quantities. Experiments have shown that the mouse does not sense the satiety factor although it is able to produce it. The *db* gene confers a beneficial advantage in the wild when food is limited because it is more efficient at converting calories into fat; however, this efficiency is a disadvantage when food is plentiful [10]. This particular mutant has provided much information relevant to obesity [9–11, 30].

Fat: This mutant arose in strain HRS. Mice accumulate abnormal deposits of fat and reach a size of 60 g by 6 months. They are hyperinsulinemic [35].

nil (neonatal intestinal lipidosis): This mutant maps to chromosome 7. Homozygotes die *in utero* at 11 days of gestation. Heterozygotes survive but adults are small in size and have reduced fertility. The intestinal wall has white patches which are fat deposits [92].

Obese: This mutant, which maps to chromosome 6, has been studied extensively [9, 10, 12, 30, 37]. Mice reach three times normal weight. Obesity is due to an increased number of fat cells [37]. Mice have a normal response to satiety factor, but they are unable to produce it in sufficient quantities.

oed (edematous): Homozygotes die shortly after birth. Plasma contains almost no lipoproteins and reduced quantities of cholesterol, triglycerides and phospholipids. Abnormalities in the skin and a bloated appearance of newborns led to the name [81].

tub (tubby): This mutant arose in C57BL/6 and maps to chromosome 7. The animals are fat and the extra weight is adipose tissue. Blood glucose is within normal limits but insulin may reach 20 times normal levels. Insulin abnormalities are measurable before obesity is apparent. The islets of Langerhans in the pancreas are enlarged and hyperactive [12].

cld (combined lipase deficiency): Mice carrying the *cld* mutation have low levels of plasma lipoprotein lipase and hepatic lipase [62, 63, 74]. The mutation is located on chromosome 17 in the *t* locus. *cld* results in death within 3 days of birth. Massive hyperlipemia, mostly chylomicronemia, is observed. The mutation is not a deletion of both enzymes because the lipases map elsewhere in the genome and because *cld* mice produce active hepatic lipase in the liver and have 2–6 times more lipoprotein lipase protein than normal mice in other tissues. The molecular basis of this mutant is unknown, but the authors suggest that it may be the lack of a protein required for posttranslational modification of lipases.

W/W^v: Another strain with altered lipase activities is *W/W^v*. Mutations at the *W* locus on chromosome 5 cause spotted pigmentation, sterility, and macrocytic anemia. Many alleles at this locus are known because of the spotted coat color. The underlying gene function is not known, but it is thought that the pigment defect and the anemia result from some defect in the progenitor cells of erythrocytes and melanocytes. Mice carrying the W^v mutation have normal levels of lipoprotein lipase and hepatic lipase in the liver but less than half normal levels of both lipases in plasma. The mutant mice also have hypertriglyceridemia and hyper-cholesterolemia. It has been suggested that these mice have a genetic defect in secretion of lipases, in transport of lipases, or in anchoring of lipases to the endothelium [28].

SAM (senescence accelerated mouse): A new strain of mice, SAM was inbred as a model of premature aging. The aged mouse had systemic amyloidoisis. Isolation of the amyloid protein and preparation of an antibody against it showed that normal mice had a closely related protein circulating in the plasma. Further characterization indicated that this normal protein was apolipoprotein A-II, a component of HDL. Apparently the SAM mouse has a mutation in the apo A-II protein itself or in its metabolism that causes its deposition as an amyloid fiber [31, 32].

Sphingolipid storage disease: A new mutation that arose in strain BALB/c may be a model of Niemann-Pick disease. The mutant mouse has a reduced ability to esterify exogenous cholesterol and accumulates unesterified cholesterol, which may reach 15 times the normal level [75].

Genetic Tools and Resources

Inbred Strains and the Genetic Map

Several hundred inbred strains of mice exist, each constructed by a minimum of 20 generations of brother-sister mating. Some inbred strains are closely related to each other and share large portions of the genome, others are very diverse genetically, and recently some have been derived from wild mice. Each strain represents a unique sampling of the natural polymorphisms found in the wild which through repeated inbreeding have become fixed in a homozygous state. Thus a survey of inbred strains is likely to reveal some of the genetic polymorphisms found in the wild mouse population. For example, a survey of inbred strains provided variants in the structure of apo A-I and apo A-II [52], the size of HDL [52], susceptibility to atherosclerosis [68], restriction fragment length polymorphisms for apo B and apo E [51, 77], and quantitative levels of lipases [4], apo A-II [18], apo E and apo B [51].

Lists of inbred strains and the holders can be found every 4 years in *Cancer Research* (1984 is the most recent), or every year in the *Mouse Newsletter*. Two books contain some history of inbred strains [21, 58] and a brief chart of the relationship of common inbreds is available in the *Jackson Laboratory Handbook*. A large number of inbred strains are available from the Jackson Laboratory, Bar Harbor, Me. in the United States and Harwell in Great Britain.

The genetic map of the mouse grows at an exponential rate. Currently at least 1,600 named loci exist and over 900 loci have been mapped. The estimated length of the diploid genome is about 1,600 cM. Mapping data are computerized by the Genetic Resources group at Jackson Laboratory. The map is updated continuously and the latest computerized version is available upon request from Drs. Muriel Davisson or Thomas Roderick, Jackson Laboratory (Bar Harbor, ME 04609). Yearly updates are published in the *Mouse Newsletter* and in the publication *Genetic Maps*. There are over 180 'anchor loci' which are the most definitely mapped loci. Because the chromosomal arrangement of genes has been preserved during mammalian evolution [43], mapping a gene in either mouse or human predicts the map position in the other species. The length of homologous chromosomal pieces between mouse and human is about 8 cM [60]. The known homologous regions are available from the Genetic Resource from Joseph Nadeau (Jackson Laboratory) or the *Mouse Newsletter*. In the field of atherosclerosis, the homology of murine and human mutants is reviewed in this volume [53].

Congenic and Coisogenic Strains

Two strains that are identical except for a difference at a specific genetic locus are coisogenic to each other. Such strains are formed when a mutation arises in an existing inbred strain. The mutation may be maintained in a homozygous state, or if it has reduced survival or reproductive ability, it may be maintained as a heterozygote. Congenic strains are formed when a variant or mutation occurring in another strain is moved into a particular inbred strain by repeated backcrossing for at least 10 generations followed by brother-sister mating to make the variant homozygous. Congenic strains are identical except for a small stretch of DNA surrounding the mutation which is less than 1% of the genome.

Congenic or coisogenic strains are particularly useful because they allow one to examine the effects of a mutation in a constant genetic background. This is a great advantage compared to studies in humans or other experimental animals where mutants are always considered against a heterogeneous genetic background. The congenic strains are also very useful for confirming a tentative map position. Since they carry a small stretch of DNA on either side of the marker, they may have several genes in addition to the gene used to construct the strain. If one has a tentative map position predicted by a conventional genetic cross or data from a recombinant inbred set, and there are congenics that differ near the suggested site, then one can test the congenic to determine if the map position was correct. The number of congenics is extensive and they cover as much as 30% of the genome.

Recombinant Inbred Strains

Recombinant inbred (RI) strains are a relatively recent tool in mouse genetics that is replacing the more laborious backcross analysis. RI strains are constructed by crossing two progenitor strains and establishing a series of new homozygous inbred strains from their progeny [1, 84]. Each new RI strain consists of a unique mixture of genes in a homozygous state derived from the two parental strains. The set of RI strains permits rapid linkage analysis because alleles of linked genes tend to be linked in the same combination as in the parents, whereas unlinked genes are randomized. Data for segregation of markers are cumulative because each RI strain is typed for a given marker once. The data are stored in a central computer and are available to other users of the RI strains. A new genetic variant can be mapped by comparing its strain distribution pattern of alleles to the strain distribution patterns of previously typed markers. Concordant distributions indicate linkage. RI strains can also be used to determine whether a phenotypic difference between two strains is due to a single gene. If the difference is due to a single gene, then each RI strain will resemble one of the two parents and no intermediate or new phenotypes will be found. RI strains can also be used to determine if two different phenotypes are determined by the same genetic factor. For example, in the analysis of *Ath-1*, the two phenotypes of HDL-cholesterol levels and susceptibility to lesion formation appeared in the same combination found in the progenitor strains in all 47 RI strains examined. This cosegregation of phenotypes indicates that the two phenotypes are determined by the same genetic factor (or by closely linked genetic factors).

Because there are several opportunities for recombination during the construction of an RI strain, the chance of recombination is about four times greater than in a genetic cross. The formulas for calculating map distances using RI strains have been published [84].

Lists of RI strains have been published [26, 84] and new RI sets are usually mentioned in the *Mouse Newsletter*. However, some RI strains are more useful than others simply because the number of strains in a set is larger or because they have been typed for many different genetic markers. When searching for new genetic variants, the best strategy is to begin by examining the strains which are the progenitors of the most useful RI sets. These are listed in table IV and the RI sets and their holders listed in table V.

Transgenic Mice

One of the most powerful tools in mouse genetics is the ability to make transgenic mice. DNA from another strain or another species can

Table IV. Progenitors of frequently used recombinant inbred strains

C57BL/6	C57L	NZB
BALB/c	SWR	129
C3H/He	AKR	SJL
DBA/2	A/J	SM

Table V. Useful recombinant inbred sets

Progenitors		Designation	Number of strains	Holder
BALB/cBy	C57BL/6By	CXB	7	Bailey, Jackson Lab.
C57BL/6J	C3H/HeJ	BXH	13	Taylor, Jackson Lab.
C57BL/6J	DBA/2J	BXD	24	Taylor, Jackson Lab.
AKR/J	C57L/J	AKXL	18	Taylor, Jackson Lab.
SWR/J	C57L/J	SWXL	7	Taylor, Jackson Lab.
AKR/J	DBA/2J	AKXD	30	Taylor, Jackson Lab.
A/JZ	C57BL/6J	AXB	45	Nesbitt, UC San Diego
NB/B1NJ	129/J	NX129	8	Taylor, Jackson Lab.
NB/B1NJ	SM/J	NXSM	19	Eicher, Jackson Lab.
BALB/cJPas	DBA/2JPas	CXD	10	Guenet, Pasteur Inst.
129/SuPas	C57BL/65Pas	129XB	15	Guenet, Pasteur Inst.
BALB/C	SJL	CXJ	12	Geckeler, Salk Inst.
SWR	SJL	SWXJ	14	Taylor, Jackson Lab.

Additional RI sets exist but are less useful because of the small number of strains per set or loci mapped.

be injected into the mouse embryo. In a fraction of such injections, the DNA is incorporated into the genome of the mouse and the gene can be studied in its new environment. This approach, which holds great promise, has been used twice in the field of lipid transport at this point in time. Transgenic mice, containing the human apo A-I gene, were used to examine the differential expression of A-I in liver and intestine [91]. The human LDL receptor gene under control of a highly efficient promotor was injected into mouse embryos and a strain with high levels of LDL receptors constructed. This strain rapidly cleared LDL from plasma and levels of apo B-100 and apo E declined, but apo A-I was not affected [33].

DNA Resource

To facilitate mapping using DNA polymorphisms, the Jackson Laboratory maintains a DNA resource. One can purchase small quantities of DNA which have been isolated from many inbred strains, mutants, and recombinant inbred strains. DNA from the large A × B recombinant inbred set is available from Dr. Muriel Nesbitt (University of California, San Diego, Calif.).

Murine Lipoproteins

Characterization of Lipoprotein Profile

Three separate laboratories have characterized murine lipoproteins employing classical ultracentrifugation to separate lipoproteins into density classes [6, 44, 57]. All three laboratories used plasma from mice fed a low fat diet, and the results are in basic agreement with each other. Camus et al. [6], using Swiss mouse fed laboratory chow, separated plasma lipoproteins into twelve density fractions and characterized each fraction for lipid and apoprotein content. The majority of lipoproteins in chow-fed mice are associated with the high density lipoprotein classes (HDL). The density fraction from $d = 1.060–1.188$ accounted for 93% of total apolipoproteins and 62% of lipids. Of the remaining lipid, 30% (by weight) was cholesterol rich (density fraction $1.033 < d < 1.060$) and 70% was triglyceride rich ($d < 1.033$). Lipoproteins in the 1.046 to 1.060 density interval had beta-mobility upon agarose electrophoresis and may correspond to the human LDL equivalent while lipoproteins of $d < 1.023$ had prebeta-mobility and are probably analogous to human VLDL. The chemical compositions of lipoproteins determined by LeBoeuf et al. [45] and Camus et al. [6] for Swiss mice and by Morrisett et al. [57] for strains CBA and C57BR were similar.

The data from these three laboratories indicate that the two major lipoprotein classes in mice fed laboratory chow are VLDL ($d < 1.04$) and HDL ($1.08 < d < 1.14$) [6]. Both of these lipoproteins are compositionally analogous to the human lipoprotein forms. VLDL is enriched in triglycerides but contains small amounts of cholesterol and cholesteryl esters. HDL ranges from 46 to 56% protein by weight depending upon the strain analyzed, and the major lipid classes are phospholipids and cholesteryl esters.

Quantitation of lipoproteins by Sudan black or oil-red-O staining of agarose gels indicates that HDL represents about 70% of total plasma lipoproteins, which agrees well with the ultracentrifugal data. The excellent agreement stems from the fact that these strains react with the neutral lipids, cholesteryl ester and triglyceride, and that murine plasma neutral lipids are mostly cholesteryl esters.

The analysis of plasma lipoproteins from mice fed atherogenic diet is less extensive. Changes in lipoprotein profile occur within 24 h after mice begin consuming an atherogenic diet, but stabilization of metabolic changes does not occur until 3 weeks [67]. After this time the total plasma cholesterol increases 2- to 3-fold in most stains of mice, reflecting the increase of lipoproteins which have beta- and prebeta-mobilities when electrophoresed in agarose. These lipoproteins constitute the major class in mice fed atherogenic diet.

Triglyceride levels usually decrease in mice fed atherogenic diets [45, 69]. Breckenridge et al. [5] indicate that triglycerides (1% of total lipid) fell to undetectable levels while the cholesterol ester content increased approximately 20%, suggesting that cholesterol ester substitutes for lipoprotein core triglyceride. Low triglyceride content of beta- and prebeta-migrating lipoproteins may be related to the relatively high levels of plasma free fatty acids, monoacyl fatty acids and glycerol which accompanies high phospholipase activity [5, 24, 83].

However, there are discrepancies in the measured triglyceride levels probably caused by the unexpectedly high free glycerol content of mouse plasma. When triglycerides are determined by the enzymatic method, which measures the glycerol moiety of triglyceride levels in the range of 100–200 mg/dl are reported [6, 57, 69] even when the authors mention that values are corrected for the free endogenous glycerol component. When triglycerides are determined by gas liquid chromatography or thin layer chromatography, much lower triglyceride levels in the range of 10–60 mg/dl are found [40, Ishida, unpubl., LeBoeuf, pers. commun.].

The levels of HDL decrease in response to the atherogenic diet in some strains but not in others. As discussed earlier, the reduction in HDL is correlated with susceptibility to the formation of atherosclerotic lesions in the aorta and coronary arteries [68].

HDL apoprotein composition appears to remain relatively unchanged in response to the atherogenic diet as indicated by SDS-PAGE in strains C57BL/6 and C3H. The major apoprotein remains apo A-I and lower amounts of apo A-II and apo C are present. Native gradient PAGE of HDL from strain C57BL/6 show only a small or insignificant change in the HDL modal peak size after consumption of atherogenic diet, and the size distribution remains symmetric [36]. These data indicate that the decrease of HDL caused by atherogenic diet is best explained by a reduction of the normal HDL pool equally affecting all HDL subpopulations. Total plasma apo E increases in response to an atherogenic diet [51, 57], and one report suggests, without much direct evidence, that an apo E-rich HDL is formed [57].

HDL

Further characterization of murine lipoproteins for size and apoprotein composition has concentrated on the HDL class. The chemical composition reported by different laboratories for HDL are in excellent agreement and resemble human HDL. In the analysis of strains C57BR and CBA [5], HDL contained an equal mass of lipid and protein. The lipid components included cholesterol ester (22%), free cholesterol (2%), phospholipid (22%), and triglyceride (1%) [5]. In the analysis of strains C57BL/6 and C3H/HeJ, HDL apoproteins were primarily apo A-I (over 90%) by scanning densitometry of Coomassie-stained alkaline-urea PAGE with smaller amounts of apo A-II, and trace amounts of apo C, and apo E [Ishida, unpul.]. A similar apoprotein distribution has been reported for the Swiss mouse [45], but Camus et al. [6] report that apo A-I constitutes only 40–60% of HDL apoproteins in Swiss mice.

The question of whether murine HDL has distinct subpopulations, analogous to human HDL_2 and HDL_3, has not been settled since data from different laboratories do not agree. Mouse HDL are fairly homogeneous when analyzed by velocity sedimentation [57], isopycnic centrifugation, nondenaturing gel electrophoresis [57], and molecular exclusion chromatography [5]. Analytical ultracentrifugation, for example, revealed a single lipoprotein fraction of flotation rate 4.2, which presumably describes the major fraction isolated in the 1.085–1.111 density interval [6]. A homogeneous HDL particle sized population is also indicated for strains CBA and C57BR by nondenaturing gradient PAGE, which results in a symmetrical narrowly sized HDL peak of approximately 100 Å (234 kd) [57]. Using the same method, homogeneously sized HDL of 102 and 95 Å were found for strains C3H and C57BL/6 respectively [36]. Thus, nondenaturing gradient PAGE data places mouse HDL near the HDL_2/HDL_3 boundary as defined for human lipoproteins [61]. This confirms the analytical ultracentrifugation results of Camus et al. [6] who reported that the predominant species of HDL sedimented with a character intermediate between HDL_2 and HDL_3.

However, other data suggest that distinct classes of HDL particles may exist. In a survey involving 40 inbred mouse strains, most strains were found to have two or more distinct HDL particle sizes as judged by nondenaturing 5% PAGE [52]. This observation is similar to human HDL, which displays a bimodal distribution in nondenaturing gradient PAGE. The apparent discordant results obtained by similar electrophoretic techniques are resolved if HDL lipoproteins are separated on the basis of both charge and size by native 5% PAGE for size discrimination is strictly observed by nondenaturing gradient PAGE when run to equilibrium. Thus, HDL of the mouse is homogeneous by size and density but may be heterogeneous with respect to intrinsic charge due to similarly sized particles varying in apoprotein composition.

LDL and VLDL

Much less is known of the murine apo B-containing lipoproteins. The most complete description is by Camus et al. [6] who found that as in the human, all apo B can be recovered in the d<1.060 interval region. This fraction accounted for about 7% of total plasma apolipoproteins and 36% of lipids. Lipoproteins analogous to VLDL and LDL are represented in the density intervals of 1.017–1.023 and 1.033–1.060, respectively. These are similar to, but not exactly the same as, the density boundaries of the corresponding human lipoproteins. Mouse LDL and VLDL were heterogeneous in size ranging from 220–280 to 310–450 Å in diameter, respectively [36]. Thus, the size of mouse and human LDL is comparable, but mouse VLDL is more polydisperse than human VLDL [61].

The major apoprotein in mouse VLDL and LDL is apo B, as expected from the data from other species. In addition to apo B, mouse LDL contains trace amounts of apo E, and mouse VLDL contains substantial quantities of apo E and apo C [36, 45]. Others have reported apo A-I in mouse VLDL and LDL [6, 22], but we think it likely that this apo A-I in due to contamination resulting from the method of lipoprotein preparation. Camus et al. [6] and Forgez et al. [22] used single isopycnic density ultracentrifugation, a method that permits lipoprotein characterization over a wide range of hydrated densities. Isopycnic ultracentrifugation has the advantage of identifying lipoprotein polydispersity, but has the disadvantage that fractions of homogeneous composition are difficult to obtain. Sequential density ultracentrifugation [29], the method used by LeBoeuf and Ishida, often provides a homogeneous fraction which is less contaminated by nonlipoprotein components of plasma. In fairness, it is noted that ultracentrifuge often dissociates apoproteins and may obscure the accurate determination of apoprotein constituents of lipoproteins obtained by either method [8, 41].

Apolipoproteins

For those apoproteins studied, mouse and human apoproteins are biochemically similar, but generally differ enough so that antibodies to either human or mouse apoproteins do not cross-react or cross-react partially with the apoprotein from the other species. Mouse apo A-I is found almost entirely associated with the HDL density fraction as determined by ultracentrifugation. The molecular weight of mouse apo A-I as determined by SDS-PAGE is 26,000 [43], 25,000–27,000 [6], or 27,000–30,000 [22], which correspond well to the molecular weight of human A-I, which is 28,000. The amino acid composition is similar to human A-I but contains a single mole each of isoleucine and cysteine which are absent in human A-I. The mouse A-I electrophoresed in alkaline PAGE often resolves as two closely migrating bands as does the human apoprotein. Isoelectric focusing detects up to five charge isomeric forms [52].

The plasma levels of apo A-I among inbred strains varied from 80 to 120 mg/dl [56, LeBoeuf, pers. commun.], values that are somewhat lower than those found in human. These low values for plasma apo A-I, however, may be biased due to the methodology used. One laboratory employed gel electrophoresis followed by densitometry of Coomassie-stained proteins [57]. The accuracy of this method suffers if the chromogenicity index (dye bound/weight protein) has not been documented. The second laboratory used Western blots (LeBoeuf), a method that may give lower values if quantitative electrophoretic transfer and recovery of proteins to membranes such as nitrocellulose did not occur [23]. Using radial immunodiffusion with mouse apo A-I as standard, values of apo A-I are 160 mg/dl, which is more comparable to human levels [Ishida, unpubl.].

The primary structure of apo A-II in mouse strain BALB/c has been determined [56]. Mouse and rat apo A-II lack cysteine which may account for its monomeric form and its size which is about half of that reported for human apo A-II. The molecular weight of mouse apo A-II has been reported as 11,000 when determined by SDS-PAGE [22] or as 8,715 when analyzed by fast atom bombardment mass spectroscopy [56]. Greater confidence in the accuracy of the latter value can be expected due to the close amino acid composition to human apo A-II which has been accurately characterized with a apparent molecular weight 8,690 [95] while the poor resolution and calibration of SDS-PAGE in this molecular weight region often lead to anomalous values. The amino acid composition is similar to the rat and human A-II, except that mouse A-II has a histidyl residue which is absent in rat and human, and arginine is present in mouse and rat but absent in the human apo A-II [22, 44, 56]. In the BALB/c strain, apo A-II comprises up to 10% of total apo HDL apoproteins by scanning densitometry of alkaline-urea polyacrylamide gels [Ishida, unpubl.].

The mouse apo A-IV gene has been cloned and sequenced and the amino acid sequence deduced [95]. As with the other mouse apoproteins, striking DNA homology was found to rat A-IV (82%); however, only 58.6% homology was found between mouse and human A-IV. Despite the lower homology for human A-IV, alignment of 14 amphipathic stretches of 22 amino acid residues were characterized for the mouse, rat and human proteins. Williams et al. [95] noticed that the amount of apo A-IV mRNA was increased 10-fold by high-fat diet for the atherosclerosis-susceptible strain C57BL/6 but unchanged in atherosclerosis-resistant strains, BALB/c and C3H.

The molecular weight of mouse apo E is 35 kd as secreted in lipoproteins from cultured mouse peritoneal macrophages. Three charge isoforms of apo E are reported with a pI of 5.4 for the major form. In vitro experiments showed that cholesterol loading of cultured macrophages stimulated the production of apo E about 4-fold [3]. An atherogenic diet increased the quantity of apo E in most inbred strains [51].

Apo B is a major apoprotein component of both VLDL and LDL. As in other species, mouse apo B occurs in two molecular forms. The nomenclature of apo B is somewhat confusing; the two forms are called B_L and B_H [82] or PI and PIII [19] or B-48 and B-100 [27]. Camus et al. [6] found VLDL (d<1.017) was enriched in apo B_L ($M_r = 250$ kd) while LDL (d = 1.033–1.060) was predominantly B_H ($M_r = 500$ kd). Similar results were described by LeBoeuf et al. [45], who reported a lower molecular weight ($M_r = 220$ and 350 kd) [19]. Recently, Lusis et al. [51] investigated the genetic control of apo B expression and reported that B-48 and B-100 were 264 and 549 kd respectively. Mouse apo B cross-reacts with antibody to human apo B [6]. The chromosomal map locations of the apo B gene in mouse and human are homologous [51]. These reports show that mouse apolipoprotein B species are closely related to other animal forms and suggest a strong genetic conservation of the apo B gene. While a direct amino acid composition of mouse apo B species has not been reported, a good interspecies correlation can be expected in light of the similar compositional analyses of apo B species between the rat [82] and human [27].

Among the known apolipoproteins, the apolipoprotein C family is poorly characterized in the mouse. The major observation regarding this apolipoprotein (observed as the most acidic apolipoprotein by alkaline urea PAGE) is the presence of a single molecular form which is distributed throughout the lipoprotein classes but primarily in the lipoproteins of density 1.033–1.060 [6]. Further characterization by Forgez et al. [22] established the similarity of murine apo C to human apo c-III by amino acid analysis, partial N-terminal amino acid sequencing, molecular weight (M_r 9,600), and isoelectric point (pI = 4.74). However, mouse apo C contains 2 mol of isoleucine which are absent in human C-III. High resolution gradient SDS-PAGE of purified apolipoproteins suggest a greater molecular weight for apo C than apo A-II [Ishida, unpubl.]

despite the opposite molecular weight relationship reported for these apolipoproteins [22].

List of Resources

Mouse Newsletter: This publication is available at US$20/year from Journals Subscription Department, Oxford University Press, Walton Street, Oxford OX2 6DP, UK. It is published three times a year: the first issue lists mutant genes and symbols, the second issue contains genetic maps, and the third issue lists inbred strains and their sources. It also contains research results, new linkages, and new mutations and chromosomal variants.

Computer Program for Maintaining a Mouse Room: Available as floppy disk for $5.00 from Dr. Lee Silver, Department of Molecular Biology, Princeton University, Princeton, NJ 08544, USA. This software package can be run on the IBM PC-XT and AT personal computers. It keeps track of mice, litters, tissues or DNA samples, and restriction digests of DNA samples.

List of Mutations and Mutant Stocks of the Mouse: This list is prepared every two years by Priscilla Lane, Jackson Laboratory, Bar Harbor, Me., USA. It lists the known genetic mutants, chromosome aberrations, and inbred strains.

Genetic Variants and Strains of the Laboratory Mouse: Edited by M. C. Green, Gustav Fischer Verlag, New York 1981. This book is a catalog of useful lists and contains a description of mutant genes, chromosomal variants, genetic maps, inbred strains, congenic strains, and recombinant inbred strains. A new edition expected in the near future edited by Mary Lyon.

Jackson Laboratory Handbook: This is available for $5.00 from the Jackson Laboratory, Bar Harbor, ME 04609, USA. It contains useful information such as the strains available for sale, the breeding performance, and relationship of common inbred strains.

References

1   Bailey, D.: Recombinant inbred strains and bilineal congenic strains; in Foster, Small, Fox, The mouse in biomedical research – History, genetics and wild mice, pp. 223–239 (Academic Press, New York 1981).
2   Ball, C.; Williams, W.; Collum, J.: Cardiovascular lesions in Swiss mice fed a high-fat-low-protein diet with and without betaine supplementation. Anat. Rec. *145:* 49–59 (1963).

3   Basu, S.; Brown, M.; Ho, Y.; Havel, R.; Goldstein, J.: Mouse macrophages synthe-
     size and secrete a protein resembling apolipoprotein E. Proc. natn. Acad. Sci. USA
     78: 7545–7549 (1981).
4   Ben-Zeev, O.; Lusis, A.; LeBoeuf, R.; Nikazy, J.; Schotz, M.: Evidence for indepen-
     dent genetic regulation of heart and adipose lipoprotein lipase activity. J. biol. Chem
     258: 13632–13636 (1983).
5   Breckenridge, W.; Roberts, A.; Kuksis, A.: Lipoprotein levels in genetically selected
     mice with increased susceptibility to atherosclerosis. Arteriosclerosis *5:* 256–264
     (1985).
6   Camus, M.-C.; Chapman, M.; Forgez, P.; Laplaud, P.: Distribution and characteriza-
     tion of the serum lipoproteins and apoproteins in the mouse, *Mus musculus.* J. Lipid
     Res. *24:* 1210–1228 (1983).
7   Chapman, M.: Comparative analysis of mammalian plasma lipoproteins. Meth. En-
     zym. *129:* 70–143 (1986).
8   Cheung, M.; Wolf, A.: Differential effect of ultracentrifugation on apoprotein A-I
     containing lipoprotein subpopulations. J. Lipid Res. *29:* 15–24 (1988).
9   Coleman, D.: Obese and diabetes: two mutant genes causing diabetes-obesity syn-
     dromes in mice. Diabetologia *14:* 141–148 (1978).
10   Coleman, D.: Obesity genes: beneficial effects in heterozygous mice. Science *203:* 663–
     665 (1979).
11   Coleman, D.: Antiobesity effects of etiocholanolones in diabetes (db), viable yellow
     (Avy), and normal mice. Endocrinology *117:* 2279–2283 (1985).
12   Coleman, D.; Eicher, E.; Southand, J.: Personal communication. Mouse Newsl. *49:*
     25 (1978).
13   Curtiss, L.; Delteer, D.; Edgington, T.: In vivo suppression of the primary immune
     response by a species of low density serum lipoprotein. J. Immun. *118:* 648–652
     (1977).
14   Davignon, J.; Gregg, L.; Sing, C.: Apolipoprotein E polymorphism and atheroscle-
     rosis. Arteriosclerosis *8:* 1–21 (1988).
15   Davis, R.; Albert, G.; Kramer, D.; Sackman, J.: Atherogenic effect of sucrose and
     white flour fed to obese mice. Experimentia *30:* 910–915 (1974).
16   Dieterich, R.; Van Pelt, R.; Glaster, W.: Diet-induced cholesterolemia and athero-
     sclerosis in wild rodents. Atherosclerosis *17:* 345–352 (1973).
17   Doering, C.; Shire, J.; Kessler, S.; Clayton, R.: Genetic and biochemical studies of the
     adrenal lipid depletion phenotype in mice. Biochem. Genet. *8:* 101–111 (1973).
18   Doolittle, M.; Lusis, A.; Lucero, J.; LeBoeuf, R.: The apolipoprotein A-II structural
     gene controls A-II synthetic rate and size of high density lipoprotein. Circulation *76:*
     IV-222 (1987).
19   Elovson, J.; Huang, Y.; Baker, N.; Kannan, R.: Apolipoprotein B is structurally and
     metabolically heterogeneous in the rat. Proc. natn. Acad. Sci. USA *78:* 157–161
     (1981).
20   Fernandes, G.; Alonso, D.; Tanaka, T.; Thaler, H.; Yunis, E.; Good, R.: Influence of
     diet on vascular lesions in autoimmune-prone B/W mice. Proc. natn. Acad. Sci. USA
     80: 874–877 (1983).
21   Festing, M.: Inbred strains in biomedical research (Macmillan Press, London
     1979).
22   Forgez, P.; Chapman, M.; Rall, S.; Camus, M.-C.: The lipid transport system in the
     mouse, *Mus musculus:* isolation and characterization of apolipoproteins B, A-I, A-II,
     and C-III, J. Lipid Res. *25:* 954–966 (1984).
23   Gershoni, J.; Palade, G.: Electrophoretic transfer of proteins from sodium

dodecylsulfate-polyacrylamide gels to a positively charged membrane filter. Analyt. Biochem. *124:* 396–405 (1982).

24 Gillett, M.; Costa, E.; Owen, J.: Evidence for phospholipase A in mouse plasma. Biochim. biophys. Acta *617:* 237–244 (1980).

25 Glueck, C.; Gartside, R.; Fullert, R.; Sielski, J.; Steinen, P.: Longevity syndromes: familial hyperalphalipoproteinemia. J. Lab. clin. Med. *88:* 941–944 (1976).

26 Green, M.: Genetic variants and strains of the laboratory mouse (Fischer, New York 1981).

27 Hardman, D.; Kane, J.: Isolation and characterization of apolipoprotein B-48. Meth. Enzym. *128:* 262–272 (1986).

28 Hatanaka, K.; Tanishita, H.; Ishibashi-Ueda, H.; Yamamoto, A.: Hyperlipidemia in mast cell-deficient W/W^v mice. Biochim. biophys. Acta *878:* 440–445 (1986).

29 Havel, R.; Eder, H.; Bragdon, J.: The distribution and chemical composition of ultracentrifugally separated human serum. J. clin. Invest. *34:* 1345–1353 (1955).

30 Herberg, L.; Coleman, D.: Laboratory animals exhibiting obesity and diabetes syndromes. Metabolism *26:* 59–99 (1977).

31 Higuchi, K.; Matsumura, A.; Hashimoto, K.; Honma, A.; Takeshita, S.; Hosokawa, M.; Yasuhira, K.; Takeda, T.: Isolation and characterization of senile amyloid-related antigenic substance (SAS$_{SAM}$) from mouse serum. J. exp. Med. *158:* 1600–1614 (1983).

32 Higuchi, K.; Yonezu, T.; Kogishi, K.; Matsumura, A.; Takeshita, S.; Higuchi, K.; Kohno, A.; Matshuchita, M.; Hosokawa, M.; Takeda, T.: Purification and characterization of a senile amyloid-related antigenic substance (apo SAS$_{SAM}$) from mouse serum. J. biol. Chem. *25:* 12834–12840 (1986).

33 Hofman, S.; Russell, D.; Brown, M.; Goldstein, J.; Hammer, R.: Over expression of low density lipoprotein (LDL) receptor eliminates LDL from plasma in transgenic mice. Science *239:* 277–281 (1988).

34 Hsu, K.-H.; Ghanta, V.; Hiramoto, R.: Immunosuppressive effect of mouse serum lipoproteins I. In vitro studies. J. Immun. *128:* 1909–1913 (1981).

35 Hummel, K.; Coleman D.: Personal communication. Mouse Newsl. *50:* 43 (1974).

36 Ishida, B.; Nichols, A.; Blanche, P.; Yashar, M.; Paigen, B.: Lipoprotein differences between the atherosclerosis-susceptible strain C57BL/6J and the atherosclerosis-resistant strain C3H/HeJ (in press, 1988).

37 Johnson, P.; Hirsch, J.: Cellularity of adipose depots in six strains of genetically obese mice. J. Lipid Res. *13:* 2–11 (1972).

38 Jokinen, M.; Clarkson, T.; Prichard, R.: Animal models in atherosclerosis research. Exp. molec. Path. *42:* 1–28 (1985).

39 Kos, W.; Lovia, R.; Snodgrass, J.; Cohen, D.; Thorpe, T.; Kaplen, A.: Inhibition of host resistance by nutritional hypercholesterolemia. Infect. Immunity *26:* 658–667 (1979).

40 Kuksis, A.; Roberts, A.; Thompson, J.; Myher, J.; Geher, K.: Plasma phosphatidylcholine/free cholesterol ratio as an indicator for atherosclerosis. Arteriosclerosis *3:* 389–397 (1983).

41 Kunitake, S.; Kane, J.: Factors affecting the integrity of high density lipoproteins in the ultracentrifuge. J. Lipid Res. *23:* 936–940 (1982).

42 Kunitomo, J.; Yamaguchi, Y.; Matsushima, K.; Bando, Y.: Cholesterol metabolism in serum and aorta of inbred mice fed a high-cholesterol diet. Jap. J. Pharmac. *34:* 153–158 (1984).

43 Lalley, P.; Minna, J.; Francke, U.: Conservation of autosomal gene syntemy groups in mouse and man. Nature *274:* 160–163 (1978).

44   Lax, S. E.; John, K. M.; Brewer, H. B., Jr.: Isolation and characterization of ApoLp-Gln-II (Apo A-II), a plasma high density apolipoprotein containing two identical polypeptide chains. J. biol. Chem. *247:* 7510–7518 (1972).
45   LeBoeuf, R.; Puppione, D.; Schumaker, V.; Lusis, A.: Genetic control of lipid transport in mice. I. Structural properties and polymorphisms of plasma lipoproteins. J. biol. Chem. *258:* 5063–5070 (1983).
46   Leiter, E.; Prochazka, M.; Coleman, D.: The nonobese diabetic (NOD) mouse. Am. J. Path. *128:* 380–383 (1987).
47   Levy, J.; Ibrahim, A.; Shirai, T.; Ohta, K.; Nagasawa, R.; Yoshida, H.; Estes, J.; Gardner, M.: Dietary fat affects immune complex disease in NZB/NZW mice. Proc. natn. Acad. Sci. USA *79:* 1974–1978 (1982).
48   Locniskar, M.; Nauss, K.; Newberne, P.: The effect of quality and quantity of dietary fat on the immune system. J. Nutr. *113:* 951–961 (1983).
49   Loria, R.; Kilbrick, S.; Downing, D.; Madge, G.; Fillios, L.: Effects of prolonged hypercholesterolemia in the mouse. Nutr. Rep. Int. *13:* 509–518 (1976).
50   Lusis, A.; LeBoeuf, R.: Genetic control of plasma lipid transport mouse model. Meth. Enzym. *128:* 877–894 (1980).
51   Lusis, A.; Taylor, B.; Quon, D.; Zollman, S.; LeBoeuf, R.: Genetic factors controlling structure and expression of apolipoproteins B and E in mice. J. biol. Chem. *262:* 7594–7604 (1987).
52   Lusis, A.; Taylor, B.; Wangenstein, R.; LeBoeuf, R.: Genetic control of lipid transport in mice. II. Genes controlling structure of high density lipoproteins. J. biol. Chem. *258:* 5071–5078 (1983).
53   Lusis, A. J.; Sparkes, R. S.: Chromosomal organization of genes involved in plasma lipoprotein metabolism: human and mouse 'fat maps'. Monogr. hum. Genet., vol. 12, pp. 79–94 (Karger, Basel 1989).
54   Mark, D.; Alonso, D.; Quimby, F.; Thaler, H.; Kim, Y.; Fernandes, G.; Good, R.; Weksler, M.: Effects of nutrition on disease and life span. I. Immune responses, cardiovascular pathology, and life span in MRL mice. Am. J. Path. *117:* 110–124 (1984).
55   Mark, D.; Alonso, D.; Tack-Goldman, K.; Thaler, H.; Tremoli, E.; Weksler, B.; Weksler, M.: Effects of nutrition on disease and life span. II. Vascular disease, serum cholesterol, serum thromboxane, and heart-produced prostacyclin in MRL mice. Am. J. Path. *117:* 125–131 (1984).
56   Miller, C.; Lee, T.; LeBoeuf, R.; Shively, J.: Primary structure of apolipoprotein A-II from inbred mouse strain BALB/c. J. Lipid Res. *28:* 311–319 (1987).
57   Morrisett, J.; Kim, H.-S.; Patsch, J.; Datta, S.; Trentin, J.: Genetic susceptibility and resistance to diet-induced atherosclerosis and hyperlipoproteinemia. Arteriosclerosis *2:* 312–324 (1982).
58   Morse, S.: Origins of inbred mice (Academic Press, New York 1978).
59   Mulvihill, B.; Walker, B.: Dietary fatty acids – cholesterol interaction and serum cholesterol in the hypercholesterolemic mouse. Nutr. Res. *4:* 611–619 (1984).
60   Nadeau, J.; Taylor, B.: Lengths of chromosomal segments conserved since divergence of man and mouse. Proc. natn. Acad. Sci. USA *81:* 814–818 (1984).
61   Nichols, A.; Krauss, R.; Musliner, T.: Nondenaturing polyacrylamide gradient electrophoresis. Meth. Enzym. *128:* 417–431 (1986).
62   Olivecrona, T.; Bengtsson-Olivecrona, G.; Chernick, S.; Scow, R.: Effect of combined lipase deficiency (cld/cld) on hepatic and lipoprotein lipase activities in liver and plasma of newborn mice. Biochim. biophys. Acta *876:* 243–248 (1986).
63   Olivecrona, T.; Chernick, S.; Bengtsson-Olivecrona, G.; Paterniti, J., Jr.; Brown, W.;

Scow, R.: Combined lipase deficiency (cld/cld) in mice. Demostration that an inactive form of lipoprotein lipase is synthesized. J. biol. Chem. *260:* 2552–2557 (1985).

64 Ordovas, J.; Schaefer, E.; Salem, D.; Ward, R.; Glueck, C.; Vorgani, C.; Wilson, P.; Karathanasis, S.: Apolipoprotein A-I gene polymorphism associated with premature coronary artery disease and familial hypoalphalipoproteinemia. New Engl. J. Med. *314:* 671–677 (1986).

65 Paigen, B.; Albee, D.; Holmes, P.; Mitchell, D.: Genetic analysis of murine strains C57BL/6J and C3H/Hej to confirm the map position of Ath-1, a gene determining atherosclerosis susceptibility. Biochem. Genet. *23:* 501–511 (1987).

66 Paigen, B.; Havens, M.; Morrow, A.: Effect of 3-methylcholanthrene on the development of aortic lesions in mice. Cancer Res. *45:* 3850–3855 (1985).

67 Paigen, B.; Holmes, P.; Mitchell, D.; Albee, D.: Comparison of atherosclerostic lesions and HDL-lipid levels in male, female, and testosterone-treated female mice from strains C57BL/6. BALB/c, and C3H. Atherosclerosis *64:* 215–221 (1987).

68 Paigen, B.; Mitchell, D.; Reue, K.; Morrow, A.; Lusis, A.; LeBoeuf, R.: *Ath-1,* a gene determining atherosclerosis susceptibility and high density lipoprotein levels in mice. Proc. natn. Acad. Sci USA *84:* 3763–3767 (1987).

69 Paigen, B.; Morrow, A.; Brandon, C.; Mitchell, D.; Holmes, P.: Variation in susceptibility to atherosclerosis among inbred strains of mice. Atherosclerosis *57:* 65–73 (1985).

70 Paigen, B.; Morrow, A.; Holmes, P.; Mitchell, D.; Williams, R: Quantitative assessment of atherosclerotic lesions in mice. Atherosclerosis *68:* 231–240 (1987).

71 Paigen, B.; Mitchell, D.; Holmes, P.; Albee, D.: Genetic analysis of strains C57BL/6J and BALB/cJ for *Ath-1,* a gene determining atherosclerosis susceptibility in mice. Biochem. Genet. *25:* 881–892 (1987).

72 Paigen, B.; Holmes, P.; Tuszyuski, G.; Switalska, H.-I.; Novak, E.; Swank, R.: Analysis of atherosclerosis susceptibility in mice with genetic defects in platelet function: correlation with platelet thrombospondin levels (in press, 1988).

73 Paigen, B.; Nesbitt, M.; Albee, D.; Mitchell, D.; LeBoeuf, R.: *Ath-2,* a second gene affecting atherosclerosis susceptibility in mice (in press, 1988).

74 Paterniti, J.; Brown, W.; Ginsbereg, H.; Artzt, K.: Combined lipase deficiency (cld): a lethal mutation on chromosome 17 of the mouse. Science *221:* 167–169 (1983).

75 Pentchev, P.; Boothe, A.; Kruth, H.; Weintroub, H.; Stivers, J.; Brady, R.: A genetic storage disorder in BALB/c mice with a metabolic block in esterification of exogenous cholesterol. J. biol. Chem. *259:* 5784–5791 (1984).

76 Prochazka, M.; Leiter, E.; Serreze, D.; Coleman, D.: Three recessive loci required for insulin-dependent diabetes in nonobese diabetic mice. Science *237:* 286–289 (1987).

77 Reue, K.; Quon, D.; O'Donnell, K.; Dizikes, G.; Fareed, G.; Lusis, A.: Cloning and regulation of messenger RNA for mouse apolipoprotein E. J. biol. Chem. *259:* 2100–2107 (1984).

78 Roberts, A.; Thompson, J.: Inbred mice and their hybrids as an animal model for atherosclerosis research. Adv. exp. Med. Biol. *67:* 313–321 (1976).

79 Roberts, A.; Thompson, J.: Genetic factors in the development of atheroma and in serum total cholesterol levels in inbred mice and their hybrids. Prog. biochem. Pharmacol., vol. 13, pp. 298–308 (Karger, Basel 1977).

80 Saito, F.: A pedigree of homozygous familial hyperalphalipoproteinemia. Metabolism *33:* 629–634 (1984).

81 Schiffman, M.; Santorineon, M.; Lewis, S.; Turchin, H.; Gluecksohn-Waelsch, S.: Lipid deficiencies, leukocytosis, brittle skin – a lethal syndrome caused by a recessive mutation, *edematous (oed),* in the mouse. Genetics *81:* 525–536 (1975).

82 Sparks, C.; Marsh, J.: Analysis of lipoprotein apoproteins by SDS-gel filtration column chromatography. J. Lipid Res. *22:* 514–518 (1981).

83 Suzuki, A.; Kawakami, M.: A hemolytic lipoprotein containing lysophosphatidylcholine produced in incubated mouse plasma. Biochim. biophys. Acta *753:* 236–243 (1983).

84 Taylor, B.: Recombinant inbred strains: use in gene mapping; in Morse, Origins of inbred mice, pp. 423–432 (Academic Press, New York 1978).

85 Taylor, B.; Meier, H.; Whitten, W.: Chromosomal location and site of action of the adrenal lipid depletion gene of the mouse. Genetics *77:* s65 (1974).

86 Thompson, J.: Atheromata in an inbred strain of mice. J. Atheroscler. Res. *10:* 113–122 (1969).

87 Verstuyft, J.; Williams, R.; Paigen, B.: Pathology of atherosclerotic lesions in mice inbred strains of mice: C57BL/6J, BALB/cJ, and C3H/HeJ (in press, 1988).

88 Vesselinovitch, D.; Wissler, R.; Doull, J.: Experimental production of atherosclerosis in mice. 1. Effect of various synthetic diets and radiation on survival time. food consumption and body weight in mice. J. Atheroscler. Res. *8:* 483–495 (1968).

89 Vesselinovitch, D.; Wissler, R.: Experimental production of atherosclerosis in mice. 2. Effects of atherogenic and high-fat diets on vascular changes in chronically and acutely irradiated mice. J. Atheroscler. Res. *8:* 497–523 (1968).

90 Walker, B.; Mulvihill, B.: Plasma cholesterol response to dietary saturated and hydrogenated fats in CBA and C57BR/cdJ mice. Nutr. Res. *4:* 601–610 (1984).

91 Walsh, A.; Breslow, J.: Cis-acting DNA elements required for apo A-I gene expression are different for liver and intestine: studies with transgenic mice. Circulation *76:* IV-222 (1987).

92 Wallace, M.; Herbertson, B.: Neonatal intestinal lipidosis in mice. An inherited disorder of the intestinal lymphatic vessels. J. med. Genet. *6:* 361–375 (1969).

93 Wallace, M.; MacSwiney, F.: Personal communication. Mouse Newsl. *58:* 20 (1975).

94 Wang, R.; Chang, S.; Kuan, C.: Experimental atherosclerosis in mice. Sci. Sininca *14:* 488–494 (1965).

95 Williams, S.; Bruckheimer, S.; Lusis, A.; LeBoeuf, R.; Kinniburgh, A.: Mouse apolipoprotein A-IV gene: nucleotide sequence and induction by high-lipid diet. Mol. cell. Biol. *6:* 3807–3814 (1986).

Beverly Paigen, PhD, Children's Hospital, 747 52nd Street, Oakland, CA 94609 (USA)

Subject Index